# FOOD POISONING PREVENTION
## 2ND EDITION

Greg Merry
F.A.I.E.H., F.R.S.H., M.N.E.H.A., M.A.W.W.A.
Director, Envirohealth
Lecturer in Food Safety, Central Queensland Institute of TAFE.
Director, Health and Environment, Rockhampton City Council

Copyright © Greg J. Merry 1989, 1997

All rights reserved.
Except under the conditions described in the
Copyright Act 1968 of Australia and subsequent amendments,
no part of this publication may be reproduced,
stored in a retrieval system, or transmitted in any form or by any means,
electronic, mechanical, photocopying, recording or otherwise,
without the prior permission of the copyright owner.

First published 1989 (reprinted six times)
Second edition published 1997 by
MACMILLAN EDUCATION AUSTRALIA PTY LTD
107 Moray Street, South Melbourne 3205

Associated companies and representatives
throughout the world

National Library of Australia
cataloguing in publication data

Merry Greg.
    Food poisoning prevention.
    2nd ed.
    Included index.
    ISBN 0 7329 4127 X.

    1. Food poisoning – Australia – Prevention. 2. Food
    handling. 3. Food contamination – Australia – Prevention.
    I. Title.
363. 19270994

Typeset in Plantin and Univers by
Typeset Gallery Malaysia Sdn. Bhd.

Printed in Malaysia

Cover and text design by Anne Stanhope
Cover image courtesy of The Photo Library/Alfred Pasieka

| DURHAM CO ARTS LIBRARIES & MUSEUMS | |
|---|---|
| 5547903 | |
| | 8.12.97 |
| 614.31 | £10.99 |
| | |

# *Foreword*

With a large proportion of the Australian family income being spent on food and an ever increasing amount of food money being spent on meals away from the home, a greater understanding of the dangers of food improperly handled, stored or prepared is needed.

Industry demands qualified tradespeople. However, the food industry is still the exception. Communities tend not to acknowledge the fact that untrained food industry personnel, unaware of the principles and values required to be upheld in the food industry, can cause death and serious hardship to those communities.

Entrepreneurial ability is all that is required to commence a food handling outlet in most areas of Australia. With the potential dangers that improperly handled food hold for the community this situation should not apply. Prior to any person conducting a food handling business or being employed in the area of food handling a basic course should be undertaken to obtain some knowledge to overcome problems which are a potential threat to the health and well-being of the community served.

*Food Poisoning Prevention* is a positive move to overcome the problems of untrained food industry personnel.

The book covers more than the basic fundamentals of preventing food poisoning with the text written in simple, practical terms for broad application. It was initially designed for use in TAFE courses but it is seen that it could complement any training system of Food Handling and Hygiene. It should be of interest not only to environmental health officers but to fast food chains, inservice food personnel, homemakers, high school students, teachers, tuckshops and libraries.

R.G. Spratt, J.P.
MAIHS, MRSH, FAIEH, FRSH
Past President
International Federation of Environmental Health

# Contents

Foreword   iii
Tables and Acknowledgements   ix
Preface   x

## 1 Introduction   *1*
Food-poisoning bacteria   *1*
The seven commandments for food handlers   *2*

## 2 Basic microbiology   *3*
Micro-organisms   *3*
  Bacteria   *3*
    Size   Shape   Cell structure   Movement   Growth
    Bacterial spore
  Infections and intoxications   *8*
  Viruses   *8*
  Fungi   *9*
    Yeasts   Moulds

## 3 Factors affecting bacterial growth   *11*
Properties of the food   *11*
  Nutrients   Moisture content   pH   Preservatives
Environmental conditions   *13*
  Temperature   Time   Oxygen   Humidity

## 4 Food preservation   *17*
Methods of preserving food   *17*
  Heat treatment
  Low-temperature treatment
  Dehydration   Salting   Use of sugar

Smoking    Use of spices    Use of acids
Vacuum sealing    Use of antibiotics
Irradiation    Use of ultraviolet light
Use of preservatives    Use of alcohol

## 5 Types of food poisoning  20
1 Natural food poisoning
2 Chemical food poisoning
3 Bacterial food poisoning
4 Viral foodborne infection
5 Mycotoxin poisoning

## 6 Bacterial food poisoning  22
The seven main causes of food poisoning  22
Events leading to food poisoning
The five main types of bacterial food poisoning  23
1 *Salmonella* infection
2 *Staphylococcus aureus* intoxication
3 *Clostridium perfringens* infection/intoxication
4 *Bacillus cereus* infection/intoxication
5 *Vibrio parahaemolyticus* infection

## 7 Personal hygiene and the food handler  33
Personal health  33
Body habits  35
Keeping hands clean  35
When hands should be washed

## 8 Hazard analysis of food  37
Potentially hazardous foods  37
Evaluating beef stew

## 9 An introduction to HACCP  44
The seven principles of HACCP  44
Principle 1 — Hazard analysis
Principle 2 — Identifying the Critical Control Points within the process
Principle 3 — Establishing the critical limits for preventative measures
Principle 4 — Establishing procedures to monitor critical control points

Principle 5 — Establishing the corrective action to be taken
Principle 6 — Establishing and maintaining records to document HACCP
Principle 7 — Verification of HACCP
The practical implementation of HACCP   *47*

## 10 Food hygiene and food handling   *49*
Purchasing food supplies   *49*
Storage of food   *50*
　Dried food   Canned food   Vegetables   Refrigerated food
　　Frozen food
Preparation of food   *54*
The cooking process   *54*
Displaying of food   *55*
Serving of food   *55*
Leftovers   *56*

## 11 Decontamination — cleaning and sanitising   *57*
　Some definitions
Principles involved   *57*
　Precleaning (preparation stage)   Cleaning (washing stage)
　　Sanitising (rinsing stage)   Drying and storage
Cleaning and sanitising program   *59*
　Implementing a cleaning and sanitising program
　Supervision and evaluation

## 12 Pest control   *64*
Common kitchen pests   *64*
Control of insects and rodents in food premises   *64*
　Rats and mice   Flies   Cockroaches
　Stored food insect pests

## 13 Kitchen design and construction   *74*
Design   *74*
　Layout   Cleanability
Construction   *75*
　Structural requirements for food premises

**14** Food poisoning in the home  *80*
  Food hygiene  *80*
    Bulk cooking   Use of leftover foods
    Leaving food at room temperature
    Transporting food   Faecal contamination
  Kitchen equipment  *81*
    The domestic refrigerator   Freezers
    The microwave oven   The kitchen
    sink   Food covers   Cutting boards
    Tea towels   Benchtops   Cleaning cloths/Scrubbing
      brushes/Scourers

**Appendix 1**   Ciguatera  *84*
**Appendix 2**   Incubation period, duration of illness
                and major symptoms of food poisoning  *86*
**Appendix 3**   Emerging pathogens causing foodborne
                illnesses  *88*
**Appendix 4**   Critical control point decision tree  *91*

Index  *92*

# Tables

Table 8.1   Foods that support bacterial growth   *39*

Table 8.2   Foods less likely to support bacterial growth or cause food poisoning   *43*

Table 11.1   Master cleaning and sanitising program   *60*

Table 11.2   Personal cleaning and sanitising program   *61*

Table 12.1   Multiplication table for insects and rodents   *65*

# Acknowledgements

The author would like to acknowledge the encouragement and assistance given to him by his wife, Gloria, in the preparation and development of this book.

The author and publisher thank the following for permission to reproduce copyright material:

Commonwealth of Australia — Department of Employment and Industrial Relations, p. 22; Schering Pty. Ltd., p. 70; the Queensland Department of Primary Industry, p. 72; CSIRO Division of Food Research, p. 81; Ciguatera — N. Gillespie and R. Lewis, *Toxic Plants and Animals: A Guide for Australia* (Appendix 1); Codex Committee on Food Hygiene (Appendix 4).

While every care has been taken to trace and acknowledge copyright, the publishers tender their apologies for any accidental infringement where copyright has proved untraceable.

# *P*reface

The first edition of *Food Poisoning Prevention* was acknowledged as a highly practical approach to food safety and was appreciated by many people involved in food preparation, from the householder to those involved in the food manufacturing industry. It provided knowledge essential for protecting against foodborne illness, and formed the basis for the development of food safety plans.

Since the first edition was published, I have produced a twin video package, 'Food Poisoning – The Choice is Yours!', which visually demonstrates the problems encountered in the food industry and shows how to overcome them. In this second edition, I have introduced the principles of HACCP (Hazard Analysis, Critical Control Points) and have expanded on the topic of the emerging pathogens affecting the food industry. The need to develop food safety (HACCP) plans is now being incorporated into food safety legislation around the world.

To allow management to create their own customised food safety plans, I have also developed 'HACCP Master', a world-first automatic evaluation computer program under HACCP that has been acclaimed worldwide. Another computer program, 'Cleaning Master' assists in the development of a cleaning and sanitising schedule under HACCP.

This textbook readily complements the other food safety material I have produced and provides a complete, integrated food safety package.

Greg Merry

# Introduction

It is the responsibility of managerial staff in the food service industry to maintain a high standard of hygiene to protect the public from food poisoning.

At present in Australia (with the exception of a few cities) food service managers are not required to undergo training and certification as they are in America. Consequently any person can enter the food industry with little knowledge of food hygiene and unintentionally serve contaminated food to their patrons. This text has been designed at a level of understanding to give the following an insight into the prevention of food poisoning:

- the homemaker
- school and college students
- inservice food handlers
- food service managers
- those wishing to enter the food industry.

Once the principles behind our food leglisation are fully understood, the need for training and certification within the food service industry becomes apparent. The public judge a food-poisoning outbreak very harshly and the establishment concerned rarely gets a second chance to open its doors. If you have ever suffered from food poisoning you will understand why.

## Food-poisoning bacteria

Food-poisoning bacteria

- do **not affect the taste, smell or appearance of the food.** There is no way to tell whether the food will cause food poisoning or not without costly testing.
- **know no bounds** and can strike people down at fetes, celebrations, charitable functions, picnics and in food establishments.
- can **severely affect** the very young, the elderly, the weak and the infirm causing rapid dehydration. Death could follow. If this occurs and you are found to have been negligent you may be charged with 'unlawful killing'.

The symptoms, incubation period and duration of illness vary with the different types of food-poisoning bacteria and we use this information in investigating a food-poisoning outbreak.

- The **symptoms** of food poisoning may include headaches, nausea, vomiting, abdominal cramps, diarrhoea, fever.
- The **incubation period** (time between eating the food and the appearance of the first symptoms) may be as short as 1 hour or as long as 72 hours.
- The **duration of illness** (time the symptoms last) may vary from 1 day to 7 days.

Food handlers are a **link** in the food **chain** that takes raw foods and:

- processes them
- transports them
- stores them
- prepares them before they are served to the public.

If there is a weak link at any point, the chain will eventually break. Similarly if there is contamination at any point along the food chain, results can be disastrous. Managers can easily see if their staff are **cleaning** successfully but they cannot *see* bacteria to check whether the premises have been effectively **sanitised**. It is for this reason that today's food service managers must have the knowledge to train their staff and need to be armed with a thermometer and bacterial swabbing kits.

With the influx of overseas visitors into Australia and Australians increasingly enjoying more leisure-time and holidays within the country, the tourism and hospitality industry has become large and fast growing. Our food industry is headed for massive expansion and unless we take steps *now* to reduce the hazards of food poisoning, these hazards will be magnified along with the development.

Abide by the seven commandments for food handlers wherever food is handled, whether at home or at work and the food-poisoning bacteria will be kept under control.

## *The seven commandments for food handlers*

1. Observe rules of good personal hygiene.
2. Don't work with food in any way while you are sick.
3. Keep food out of the danger zone for food poisoning during preparation, storage, serving and transportation of food.
4. Only purchase food from reputable authorised food sources.
5. Observe rules of good food hygiene practices.
6. Maintain your premises and equipment in a sound, clean and sanitary condition.
7. Prevent contamination by insects and rodents by eliminating any infestation.

# Basic microbiology

## Micro-organisms

A micro-organism is an organism that is so small that it can only be seen under a microscope. Bacteria, viruses and some fungi are all micro-organisms. These are commonly referred to as *germs*. It is a misconception to think that all micro-organisms are harmful, as less than 1% are harmful to humans. These are said to be **pathogenic.**

Micro-organisms occur naturally in our environment. Our bodies are covered with micro-organisms and we eat and breathe them in every day of our life.

We take advantage of the harmless micro-organisms' good characteristics by using them to:

- decompose leaves and garbage
- treat sewage
- produce antibiotics
- manufacture foods such as cheese, wine, beer, vinegar and yoghurt.

We even, in our testing procedures, use the presence of certain harmless bacteria to indicate if our food and water is contaminated. These micro-organisms are called coliforms and are found naturally in the intestines of warm-blooded animals, including man. If they show up in food or water, it indicates that there has been some form of contamination and that other harmful intestinal bacteria may be present.

Similarly some bacteria may be harmless to humans yet still spoil our food. Where these spoilage bacteria occur, they indicate conditions that may also be suitable for the growth of food-poisoning bacteria.

Micro-organisms can be classified into the following categories:

- bacteria
- viruses
- fungi — yeasts, moulds.

## Bacteria

### Size

Bacteria are extremely small and cannot be seen by the naked eye. They vary in size but generally measure around 1 micron (one millionth of a metre).

## Shape

Bacteria are classified by their shape.

*Coccus* (plural = cocci)
spherical

*Bacillus* (plural = bacilli)
rod-shaped

*Vibrio* (plural = vibrios)
comma-shaped

*Spirillum* (plural = spirilla)
spiral-shaped

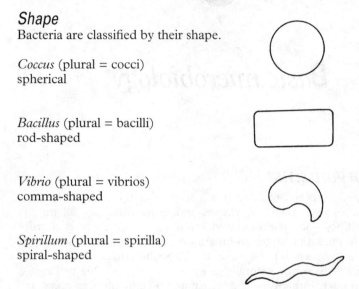

The first three are the most common shapes of bacteria found in food. The words indicating the shape of the bacteria often appear in their names.
  *Staphylococcus aureus* — Golden Staph
  *Streptococcus pyogenes* — causes Scarlet Fever
  *Bacillus cereus*
  *Vibro parahaemolyticus*

## Cell structure

Regardless of their shape, bacteria have the same internal structure.

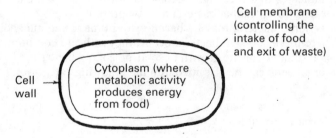

**Figure 2.1** A bacterial cell

The interior of the structure is **cytoplasm** where the bacterial food is digested. It is contained within the **cell membrane** which controls the intake of food and exit of waste. The **cell wall** gives the bacteria its characteristic shape.

## Movement

Most bacteria have no means of locomotion and have to rely on agents such as wind, water, dust, insects, rodents and humans in order to '**hitchhike**' from one place to another. Our hands are the greatest transport of bacteria to food or surfaces.

Some bacilli, for example *Salmonella*, have tail-like appendages called flagella (singular = flagellum), whose whip-like movements enable the bacteria to move in a liquid medium.

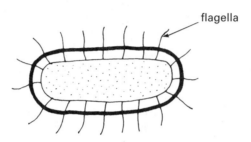

**Figure 2.2** Flagella

## Growth

Bacterial growth relates to the increase in the **number** of bacteria, not in their size. Under ideal conditions, bacteria reproduce by doubling every 15 to 20 minutes. The cell develops to its full size then divides into two. These two then develop and divide into 4 then 8, 16, 32, 64 and so on.

| | | |
|---|---|---|
| 1 cell | 1 bacterium | Start |
| | 2 bacteria | 15 mins |
| | 4 bacteria | 30 mins |
| | 8 bacteria | 45 mins |
| | 16 bacteria | 1 hour |
| | 32 bacteria | 1 hour 15 mins |
| | 64 bacteria | 1 hour 30 mins |
| | 128 bacteria | 1 hour 45 mins |
| 2 cells | 256 bacteria | 2 hours |

**Figure 2.3** Cell division

Cells can only reproduce while in the vegetative, or live, form. Bacteria multiply readily in many foods. They quickly reach the **infective dose**, which is the number of micro-organisms per gram of food necessary to produce food poisoning. Usually this is about 1 million per gram of food.

The time taken to reach the infective dose depends on the degree of contamination and factors affecting bacterial growth (see chapter 3), but is generally regarded as approximately 4–6 hours. Bacteria are transferred to a new surface or medium when, for example, food handlers scratch their nose while preparing a ham sandwich. Mucus from the nose is transferred via the hands to the slice of ham. Not just one micro-organism is added but thousands per gram of food.

## Growth phases

Under ideal conditions four phases of growth occur.

1 **Lag phase** during which the bacteria adapt to their new environment. Some may even die. This phase may be as short as 1–2 hours or as long as several days and is **the unknown factor**.
2 **Log phase** during which the bacteria, once adapted, multiply by doubling in number every 15 to 20 minutes. It is during this stage that the 'infective dose' is reached. It would take only about 2 hours for the numbers to go from 5000 to 1 000 000.
3 **Stationary phase** during which bacterial growth levels out because the bacteria are competing for available food. By this stage the ham on the sandwich would be in a state of decomposition and inedible.
4 **Death phase** in which the food supply is exhausted and waste products accumulate. Bacteria begin to die rapidly.

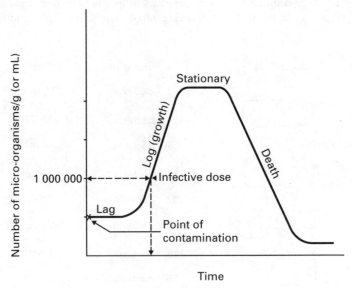

**Figure 2.4** Theoretical bacterial growth curve

Let us assume the ham is initially contaminated with 5000 micro-organisms/g of ham and the lag phase is 3 hours. Following this phase the bacterial growth in the log phase would be:

| | |
|---|---|
| 5000 | Start |
| 10 000 | 15 minutes |
| 20 000 | 30 minutes |
| 40 000 | 45 minutes |
| 80 000 | 1 hour |
| 160 000 | 1 hour 15 minutes |
| 320 000 | 1 hour 30 minutes |
| 640 000 | 1 hour 45 minutes |
| 1 280 000 | 2 hours |

In this situation only 5 hours are required to reach the infective dose. If the ham sandwich were contaminated at 7 a.m. the infective dose would be reached by 12 noon. If after this the sandwich is eaten, food poisoning could occur.

## Bacterial spores

It is generally thought that heat kills all bacteria. However, some species, for example *Bacillus* and *Clostridium* spp, have a **survival mechanism** to protect them when subjected to adverse conditions such as high and low temperatures, drying and disinfectants, over long periods of time.

When conditions become unfavourable a spore develops within the 'vegetative' (or live) cell by the formation of a special thickened wall that can protect it when the parent cell disintegrates.

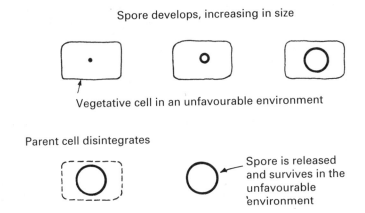

Figure 2.5 Spore development in unfavourable conditions

Spores *cannot* reproduce or increase in number but simply 'hibernate'. When favourable conditions return the spore changes back to the 'vegetative' or live cell, which is then able to reproduce.

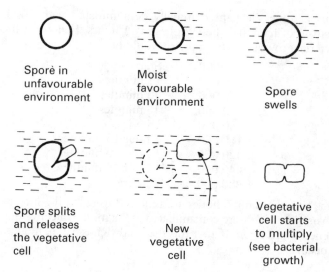

**Figure 2.6** Spore development in favourable conditions

## Infections and intoxications

A foodborne **infection** is due to the consumption of a certain **number** (infective dose) of harmful, living **micro-organisms** in food. An invasion by an infective dose of bacteria causes a reaction in the body.

A foodborne **intoxication** is due to the consumption of a **toxin** produced by certain bacteria. These bacteria upon reaching the infective dose secrete a toxin (or poison) into the food. In this case, it is the toxin that causes a reaction in the human body, *not* the bacteria themselves.

Most toxins are heat-stable. Once they are formed, heating the food may kill the bacteria, but the toxins will remain unchanged and will cause food poisoning. Some toxins are released in the food before it is eaten and some are released when the food reaches the intestine. (See infections and intoxications, chapter 6.)

## Viruses

Viruses are the smallest micro-organisms, being one-third to one-hundredth the size of an average bacterium.

These very simple structures can only reproduce inside the living cells of their host. They therefore cannot reproduce in food but merely use it as a form of transport from one person to another. To be effectively transmitted, the viruses in the food have to be present in sufficient numbers to survive the environmental conditions outside the host and still remain infective when the food is eaten. Viruses transmitted from faeces to mouth (the faecal-oral route) such as those causing hepatitis A and epidemic viral gastroenteritis (Norwalk Viruses) tend to survive adverse conditions outside the host. These are more likely to cause foodborne infections.

Some viruses such as influenza viruses are easily killed by heat, whereas others can survive temperatures up to 80°C. Viral transmission through food is a subject still under investigation. Affected foods indicate an intimate handling by the food handler and consumption without prior or further cooking.

Food vehicles include uncooked foods such as salads, fruits, sandwiches, milk drinks and other beverages, and precooked foods such as meats, pastries and milk products.

> **The seven commandments for food handlers if observed will also control viral transmission.**

# Fungi

Fungi can be divided into two groups — yeasts and moulds. Similar to bacteria there are helpful and harmful yeasts and moulds.

### Yeasts

Special cultivated yeasts are used in the making of bread and the fermentation of beer.

Wild yeast can contaminate causing food spoilage, as seen by the presence of bubbles, alcoholic smell and taste and sometimes a slime formation on fruit juices and vinegar products. Yeast cells can be destroyed by heating above 60°C for 15 minutes.

Yeasts reproduce by a process known as 'budding', where a small bulge appears and grows on the side of the mother cell. On reaching half the size of the mother cell, the daughter cell is cut off and released to grow to its full size and reproduce again.

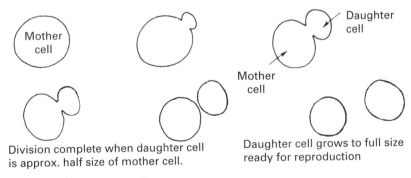

**Figure 2.7** Yeast reproduction

### Moulds

These can be seen as a group of fuzzy 'cotton-like' masses on food such as bread, cheese and citrus fruits. We do, however, use particular moulds in the making of some antibiotics and blue vein cheeses.

Moulds cause spoilage but some mould found in mouldy grain and nuts produce **mycotoxins** (see chapter 5). Most moulds can be destroyed by heating above 60°C for 10 minutes.

Moulds reproduce by means of spores, which are not the same as bacterial spores. The spores form like seeds of a flower head and when released each spore (seed) is capable of producing a new plant.

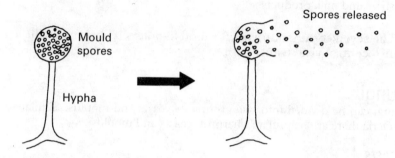

**Figure 2.8** Mould reproduction

# ▼ 3

# *F*actors affecting bacterial growth

Bacterial food poisoning can only occur if:
- food is contaminated by bacteria
- these bacteria are allowed sufficient time to multiply
- they multiply until they reach the infective dose
- the food containing the bacteria or their toxin is ingested
- the infective dose causes an infection or intoxication.

The more favourable the conditions the more the living organisms multiply. Bacterial growth depends on:
- the food having certain properties
- the environmental conditions the bacteria are subjected to.

Bacteria, in fact, require almost the same food and environmental conditions we enjoy. However, bacteria are even more adaptable than the human body.

## *P*roperties of the food

### Nutrients
Bacteria basically love the high-protein perishable foods — meat, poultry, fish, seafood, dairy products (milk, butter, cream and cheese), eggs, and all their by-products. We call these foods **potentially hazardous foods**. (Chapter 8 deals with these foods in detail.)

### Moisture content (or water activity, Aw)
Water activity is an index to indicate the amount of moisture in a food. The lower the Aw index figure, the safer the product and the longer its shelf life. The higher the index figure, the greater the growth potential for bacteria.

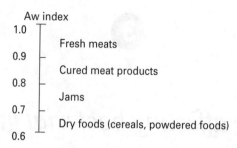

**Figure 3.1** Aw measurement

Dried products, for example powdered milk, dried egg powders, do not allow bacterial growth and need not be refrigerated. However once we add moisture or reconstitute the product, it has to be regarded as the *fresh* potentially hazardous food and refrigerated.

Spores may survive in dried food products, for example *Bacillus cereus* spores in dry custard powder. However, when the custard powder is added to the hot milk the spores 'vegetate' once the custard cools down sufficiently (into the danger zone).

Bacteria may not grow in dried food but may merely survive. Only after baby's milk formula has been reconstituted will *Salmonella* in it begin to grow again.

# pH

pH is a measurement of acidity or alkalinity. Among food products vinegar is acidic and baking powder is alkaline. Food-poisoning bacteria prefer a pH of 4.5 to 7.0, that is, slightly acidic to neutral-type foods. In foods with a pH of less than 4.5 very little bacterial growth occurs. However, moulds may grow and thrive on the very acidic type foods such as tomatoes or fruit juices, whose low pH restricts bacterial growth.

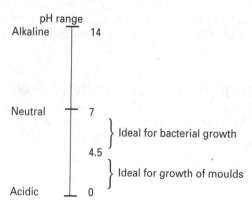

**Figure 3.2** pH measurement

## Preservatives

Under the National Health and Medical Research Council's Food Standards Code, preservatives may be added to certain foods to inhibit bacterial growth. Only specific additives are allowed for certain foods and then only in amounts up to a maximum prescribed level.

## *Environmental conditions*

## Temperature

Heat may destroy bacteria but freezing does not; it only stops their growth. Temperature is one of the best tools we have to control bacterial growth.

> The simple rule is **keep food hot** (over 60°C)
> or **keep it cold** (below 5°C)
> or **don't keep it at all.**

The temperature zone between 5°C and 60°C is known as the **danger zone** for food-poisoning bacteria as this is the temperature range in which the bacteria will multiple. Very rapid growth occurs between 20°C and 45°C and this is known as the **high danger zone.**

Now consider the temperatures of your kitchen. In summer or winter, air temperatures in the majority of cities or towns remain well within the range 5 to 60°C. Most summer kitchen temperatures are in the high danger zone, 20 to 45°C. In your own interests, you should find out your seasonal minimum and maximum air temperatures to see if they are between 5 and 60°C.

To keep food out of the danger zone store it in a refrigerator or cold room (below 5°C) or in a heated display unit or oven (above 60°C) or work *quickly* with food at room temperature. Refrigeration and freezing merely stop bacterial growth. So if food is slightly contaminated when it is placed in a refrigerator or freezer it will still be slightly contaminated when it comes out.

When we heat food above 60°C, the vegetative bacteria start to die slowly, then very rapidly as the food approaches boiling point. Spores, however, may survive even after several hours of boiling, as for example in the making of soups, stews and stocks.

# 14 Food Poisoning Prevention

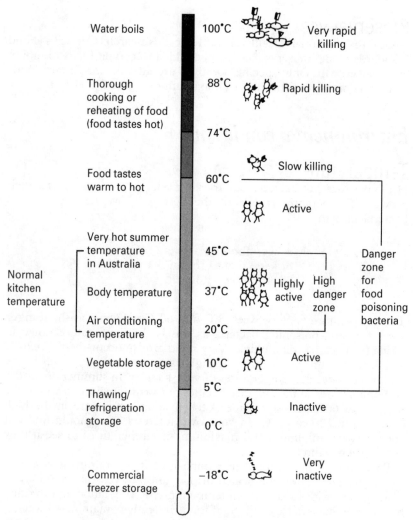

**Figure 3.3** Temperature control and bacterial growth

## Time

Food-poisoning bacteria act like a time bomb. Contamination can be likened to the setting of the bomb. There are now only 4 to 6 hours in the danger zone (5 to 60°C) before the food-poisoning bacteria reach the infective dose (the magical million mark). If the food is eaten after the infective dose is reached, the bomb goes off.

# Factors affecting bacterial growth

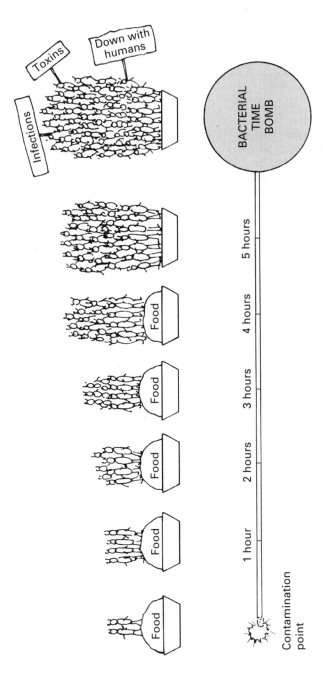

**Figure 3.4** The bacterial time bomb

> In the food industry you have to assume the food you purchase may already be contaminated.

If the food is only slightly contaminated don't allow the bacteria to double every 15 to 20 minutes by keeping it in the danger zone any longer than you have to, otherwise you will be left holding the bomb.

## Oxygen

Food-poisoning bacteria vary in their oxygen (air) requirements. Some bacteria require oxygen to grow. These are called **aerobes.** Others grow in the absence of oxygen and are called **anaerobes.** Those that can grow with or without oxygen are known as **facultative** bacteria. We use this information to advantage, by vacuum packing meats or canning food, processes in which oxygen is removed to reduce aerobic growth.

When we mince a piece of steak up, the mince has a large surface area that is in contact with numerous air pockets, therefore we increase aerobic growth. Similarly if we make a malted milk, the machine folds in and encapsulates air pockets again increasing aerobic growth.

## Humidity

When food is in a high humidity environment, the food absorbs moisture from the air, increasing its water activity and therefore its bacterial growth. When food is in a low humidity environment the air extracts moisture from the food, lowering the water activity of the surface of the food and decreasing bacterial growth.

# Food preservation

We will now look at how the factors that affect bacterial growth can be utilised in current food preservation methods.

## Methods of preserving food

### Heat treatment

This is the most commonly used method of preservation. The higher the temperature, the shorter the contact time needed. Contact time is the time micro-organisms are subjected to various sanitising or sterilising processes to cause their effective destruction. Pasteurising milk is an example of the use of this. There are three different methods of pasteurisation.

- LTLT (low temperature/long time)   Milk is heated to 61 to 65°C, held at this temperature for 30 minutes then quickly cooled to below 5°C.
- HTST (high temperature/short time)   Milk is heated to 72 to 73.5°C held there for 15 seconds then quickly cooled to below 5°C.
- UHT (ultra heat treated)   Milk is heated to not less than 133°C then aseptically packed in hermetically sealed packages. No refrigeration is necessary.

### Low-temperature treatment

- refrigeration below 5°C
- freezing below −18°C

### Dehydration

Dehydration lowers the water activity and therefore restricts bacterial growth. Foods preserved this way include powdered milk, dried egg powder, dried fruits.

### Salting

Salting is usually used in conjunction with dehydration. Salt destroys the bacteria's protoplasm. Salt is a commonly occurring preservative.

Note how often sugar and salt appear on the list of ingredients on food labels.

## Use of sugar
Sugar, used in jams, jellies etc., helps preserve them. Products with high sugar contents seldom cause food-poisoning problems as they are generally prepared using heat treatment as well.

## Smoking
Smoking is usually used in conjunction with salting and drying. The substances in the smoke have germicidal action, but their effects don't go deep. Fish, meat and hams can be preserved by smoking.

## Use of spices
In the early days before there was cold storage, spices were used to mask the taste of spoiled food. Cinnamon, cloves and mustard have very slight antibacterial powers.

## Use of acids
Vinegar (or acetic acid) is an acid and lowers the pH thereby restricting bacterial growth. Vinegar is used in pickles and chutneys.

## Vacuum sealing
Vacuum sealing by canning, bottling or cryovac packing etc. exhausts the air from the food container thus limiting the oxygen available to the aerobic bacteria. Usually it is used in conjunction with heat treatment.

## Use of antibiotics
Nisin (food additive no. 234) is an antibiotic substance permitted in canned fruits with a pH below 4.5, and in cheese.

## Irradiation
Irradiating foods with cobalt 60 is being considered for the following purposes:
- to limit insect infestations in foods such as rice
- to decontaminate spices
- to extend the shelf life of fruit and vegetables, for example mangoes and mushrooms.

Up to 1 kilogray will kill most insect pests (small dose).
1 to 20 kilogray will destroy most bacteria (medium dose).

## Use of ultraviolet light
This purple light is often seen in refrigerated display cases in butchers' shops and delicatessens. UV light can kill bacteria on surfaces exposed to this light but spores may be resistant to it.

## Use of preservatives
Preservatives are only permitted to be used in certain foods to prescribed levels. Some salamis are not heat treated and great reliance is placed on preservatives to control bacterial growth in such foods.

## Use of alcohol
Alcohol restricts the growth of micro-organisms.

# Types of food poisoning

There are various forms of food poisoning that occur in the community. They can be broadly categorised into five types.

## 1 Natural food poisoning
Various plant materials and fish are known to be toxic to man. These include:

- rhubarb leaves and spinach water, which contain oxalic acid
- 'green' potatoes which contain solamin toxin
- poisonous mushrooms
- zucchini and melons which contain curcurbitacin E
- large predatory reef fish, for example Spanish mackerel, snapper, coral trout, red emperor and coral cod which can accumulate ciguatoxin in their body. (See Appendix 1.)
- varieties of poisonous fish for example puffer fish which produce tetrodotoxin.

## 2 Chemical food poisoning
Chemicals causing food poisoning usually get into food accidentally or through negligence. Symptoms usually appear within 1 hour. Ways in which chemical contamination occurs include:

- overuse and casual use of insecticides, cleaning agents and disinfectants contaminating food and/or surfaces where food is prepared
- using storage containers which react with, or release toxic metals into, the food
- heavy metals can be found in food such as fish caught in waters receiving industrial waste (mercury and zinc)
- sodium hydroxide (caustic soda) can enter food because of inadequate rinsing of bottles cleaned with caustic soda
- monosodium glutamate (MSG, food additive no. 621) is used as a flavour enhancer in Chinese restaurants. Excessive amounts can be harmful, and can cause what is known as 'Chinese Restaurant Syndrome'.

## 3 Bacterial food poisoning

The most common bacteria causing food poisoning are *Staph. aureus*, *Salmonella*, *Clostridium perfringens*, *Bacillus cereus*, and *Vibrio parahaemolyticus*. These are dealt with in detail in chapter 6.

## 4 Viral foodborne infection

Viruses don't multiply in food. They only use food as a transport medium to gain entry into the human body where they multiply. Only a small infective dose is required. Viruses cause hepatitis A and epidemic viral gastroenteritis (Norwalk Virus).

## 5 Mycotoxin poisoning

As with bacteria there are good moulds (for example that used to make blue vein cheese) and bad moulds, some of which are capable of producing toxins, called mycotoxins.

Because of our standard of living, we do not commonly eat mouldy food. However sometimes we do unknowingly. One of the mycotoxins, aflotoxin, is produced by the mould *Aspergillus flavus*, that grows mainly in damp wheat, maize and peanuts before and after harvesting. Peanuts are now passed through modern detecting and sorting machines to remove the mouldy ones. If a batch containing one mouldy peanut is made into peanut paste, high aflotoxin levels may be found in small sections or samples of the paste and can cause poisoning. The current maximum level for total aflotoxin in peanuts is 15 µg/kg (15 parts per billion). Difficulty is experienced with techniques capable of detecting this low level.

# 6

# *B*acterial food poisoning

This is the major type of food poisoning in Australia and can be easily avoided by following the **seven commandments for food handlers** (see chapter 1). These are based upon the following.

## *The seven main causes of food poisoning*

1. Poor personal hygiene of food handlers
2. Recent intestinal or respiratory upsets of food handlers
3. Inadequate temperature control during the transport, preparation storage or serving of food
4. Food from an insanitary source of supply
5. Bad food handling practices
6. Dirty food premises and/or equipment
7. Presence of insect and rodent pests.

## Events leading to food poisoning

For bacterial food poisoning to happen, the following events and conditions need to occur in the order given.

- **Initial contamination** of the food with the appropriate bacteria.
- The food must be able to supply all the **nutrients** that the bacteria require for growth, for example meat, poultry, fish seafood, eggs, dairy products (milk, butter, cream, cheese), and their by-products.
- The **temperature** of the food needs to be in the danger zone (5 to 60°C).
- The food has to be in the danger zone for sufficient **time**, generally 4 to 6 hours, to reach the **infective dose** (approx. 1 million micro-organisms per gram of food).
- **Consumption** of that food by a human.

If you, as staff manager or handler, have neglected the seven commandments for food handlers incorporated into our food legislation in Australia and believe that you haven't caused food poisoning so far, then you have been **lucky**.

Either:
- the food was consumed *before* the bacteria reached the infective dose and/or had a chance to produce toxins, and it therefore did not cause any problems or
- the bacteria reached the infective dose but the food was left over at the end of the day and not sold or eaten or
- you did cause food poisoning and wondered why your customers never came back.

## The five main types of bacterial food poisoning

1. *Salmonella* — causes an **infection** when food containing the living bacteria is eaten.
2. *Staphylococcus aureus* — causes an **intoxication** when toxin in the food is consumed. The bacteria may be destroyed by heating the food but the toxin is heat-stable and will still cause food poisoning.
3. *Clostridium perfringens* — causes both an **infection** as a result of the consumption of living bacteria in the food and an **intoxication** due to the release of toxin by the bacteria in the intestine.
4. *Bacillus cereus* — 2 types
   (a) *vomiting type* (similar to *Staph. aureus*) **intoxication** when food containing the toxin is consumed
   (b) *diarrhoea type* (similar to *Clostridium perfringens*). This type causes both an **infection** due to consumption of living bacteria and an **intoxication** due to the release of toxin by the bacteria in the intestine.
5. *Vibrio parahaemolyticus* — causes an **infection** when living bacteria in the food are consumed.

Emerging pathogens causing foodborne illnesses and diseases are included in Appendixes 2 and 3.

## 1 *Salmonella* infection

### Causative agent
There are at least 1600 serotypes (strains) of the Salmonella group. One of the most common is *Salmonella typhimurium*.

### Description
Short thin bacilli which have flagella and are motile (with some exceptions).

### Habitat
Found in the intestine of warm-blooded animals (including humans, cattle, poultry, birds, dogs, cats, rats and mice), insects (flies and cockroaches) and their faeces. Healthy humans can be carriers.

## Growth characteristics

*Temperature*
5–45°C (optimum 37°C)

*Oxygen*
facultative

*Moisture (Aw)*
0.94

*pH*
4.5 to 9.0 (optimum 6.5 to 7.5)

## Special characteristics
Destroyed by heat but can survive drying or freezing.

## Foods involved

*Food animals*
Overcrowding and stress in food animals increase the rate of *Salmonella* in excretions, as in intensive farming. During the slaughtering and dressing process, low numbers of *Salmonella* may contaminate the carcase. Poultry is more of a problem because the small size of the chest cavity increases the chance of splitting the intestine during evisceration.

*Eggs*
*Salmonella* penetrate egg shells when they are covered with faecal matter and either cracked or moist. Ducks eggs are most susceptible to *Salmonella* contamination because ducks' ovaries may be infected by *Salmonella* organisms and because ducks lay their eggs in moist environments.

*Raw milks*
Through faecal contamination during milking, but pasteurisation kills *Salmonella*.

*Oysters*
Where 'cleansing' water is contaminated by sewage.

*Vegetables/Salads*
Act as a vehicle when grown in animal manures and not washed thoroughly. Cause cross-contamination when made up with meat or other susceptible foods.

## Mode of transmission

*Humans*
Poor personal hygiene. Faecal contamination of hands can be transferred to the food or surfaces in contact with food. Convalescent carriers can

excrete *Salmonella* in their faeces up to 12 months after they are apparently over the illness. Healthy carriers also exist.

*Cross-contamination*
- From food to food, for example raw food to cooked food.
- Surface to food or vice versa, for example a cutting board used for various different foods. Sufficient cutting boards should be available for each separate type of food or the board should be cleaned and sanitised before re-use.

## Control measures
1. Do **not** thaw frozen food at **room temperature**. Remember the danger zone. Let it thaw out thoroughly in a refrigerator or microwave oven.
2. Thoroughly cook food especially large cuts of meat, for example rolled roasts and turkeys, to kill the bacteria in the centre of the meat. **Heat alone will kill *Salmonella*.**
3. Prevent cross-contamination from raw to cooked food by using separate equipment or surfaces for each.
4. Clean and sanitise all equipment and surfaces after use. A few crumbs may support a bacterial 'time bomb'.
5. Keep food hot (above 60°C) or cold (below 5°C) and ensure foods are stored separately to prevent cross-contamination.
6. Maintain good personal hygiene. Always wash hands after going to the toilet (as the bacteria originate in the intestine) or before and after handling raw produce.

# A case history
A food handler in a barbecue chicken establishment was a healthy carrier of *Salmonella virchow*. She made large batches of stuffing for the chickens by breaking up and mixing the ingredients with her hands. This process contaminated the stuffing.

The stuffed chickens were cooked in batches on the rôtisserie but due to overcrowding, some larger chickens were not thoroughly cooked. The fact that the chicken and stuffing were only partially cooked stimulated the growth of the bacteria in the stuffing. The chickens were placed on a large stainless steel tray that was previously used for the packing of raw chickens. Here further cross-contamination of the chicken carcase occurred. The chickens were then placed in a bain marie to keep warm until sold. (Tests later showed that some chickens and stuffing were contaminated with *Salmonella*.)

The next day, unsold chickens were cut up by the *Salmonella* carrier. Chickens, stuffing and possibly the carrier's hands contaminated the cutting board. The chicken meat was used as a filling in sandwiches which were wrapped and kept at room temperature until sold, ensuring continued growth of the *Salmonella*.

It was common practice for establishment staff to wash their hands in the sink, nibble from the food during preparation and to use a towelling cloth to wipe chicken grease off their hands instead of washing their hands. The towelling cloth and sink showed contamination with *Salmonella*.

Two other staff became *Salmonella* carriers and another staff member contracted *Salmonella* either because of nibbling contaminated food or using the contaminated towelling cloth.

Had the control measures listed been enforced it would have prevented 72 cases of *Salmonella* poisoning, one of which ended in death.

## 2 *Staphylococcus aureus* intoxication

### Causative organism
*Staphylococcus aureus*

### Description
Small rounded cocci clumped together like a bunch of grapes.

### Habitat
- Normally found on a healthy human body — in the nose, throat, hair, beard, armpits, groin, ears, sleeper-type earrings, without any detrimental effect on the body. It is also found in pimples, boils, cuts, burns and abrasions.
- Udders of cows infected by mastitis give *Staph. aureus* in raw milk.

### Growth characteristics

*Temperature*
5–46°C (optimum 37°C)

*Oxygen*
facultative but prefer presence of oxygen.

*Moisture (Aw)*
0.86 (aerobic conditions)
0.9 (anaerobic conditions)

*pH*
4.2 to 9.3 (optimum 7.0 to 7.5)

### Special characteristics
1. Can survive drying and freezing.
2. Poor competitor when present with other bacteria.
3. Tolerant of high concentrations of salt (10 to 20%) which would restrict the growth of other bacteria, eliminating competitors.
4. After reaching 1 million/g of food, the bacteria produce a heat-stable toxin in the food. However temperatures above 65°C kill the bacteria.
5. Can grow in curing solutions and will tolerate 50 to 60% sugar (sucrose) content.

## Food involved
Salted ham and meat products, chicken, cold meats, custards, pastry fillings, potato salad.

## Mode of transmission
- Poor personal hygiene — usually hands touching various parts of the body harbouring *Staph. aureus* then contaminating food or food preparation surfaces. Coughing and sneezing also transfer the bacteria to food.
- Cross-contamination especially involving cutting boards.

## Control measures
1. Good personal hygiene. Make sure hands are washed, as hands are the biggest contaminator.
2. Refrain from touching the body during food handling.
3. Don't smoke in a food establishment. Saliva from the mouth is transferred onto the butt and fingers during smoking.
4. Keep food hot (above 65°C) or cold (below 5°C).
5. Clean and sanitise all equipment.
6. Refrain from unnecessary handling of food. Use utensils and/or gloves.

# A case history
A food handler suffering from a heavy cold prepared a variety of sandwiches at a canteen at 7 a.m. to be sold from 11 a.m. to 2 p.m. The fillings were placed on the bread by hand and during the preparation the food handler rubbed her running nose on the back of her hand. Large numbers of *Staph. aureus* were passed via the hand to the fillings. The sandwiches were covered and left in the display case of a hot kitchen during a tropical summer day until sold. Large bacterial numbers required only a short time to reach the infective dose and then produced the toxin.

Food handlers suffering from respiratory disorders, should not be allowed to handle food.

Better personal hygiene (that is, controlling body habits) and the use of utensils or gloves to handle food would have improved the situation.

The use of control measures again would have prevented the hospitalisation of numerous food-poisoning victims.

# 3 *Clostridium perfringens* infection/intoxication

## Causative organism
*Clostridium perfringens*

## Description
Rod-shaped bacilli, non-motile

## Habitat
- Found in intestine of animals, birds, insects and humans.
- Found as a spore in soil, dust, air and water.
- Found in vegetables containing animal faeces or dust.

## Growth characteristics

*Temperature*
15–50°C (optimum 45°C)

*Oxygen*
anaerobic

*Moisture (Aw)*
0.97

*pH*
5.0 to 9.0 (optimum 6.5 to 7.2)

## Special characteristics
1. The bacteria are killed by heat.
2. They produce spores which are heat-resistant and survive up to 4 hours boiling.
3. They produce a toxin in the intestine.
4. Infective dose = 1 million micro-organisms/g of food.
5. Growth inhibited by 5% salt.
6. They are gas-producers.

## Foods involved
- Large volumes of stew-type foods that are prepared for reheating at a later stage, for example when catering for large numbers of people at Christmas parties, weddings etc. where heating facilities and/or refrigeration are inadequate. Bacteria may be present in meat; bacteria and/or spores may be present in vegetables.
- Larger cuts of meat, for example roasts, turkeys etc.

## Mode of transmission
- Human — bad personal hygiene especially failure to wash hands. Bacteria on hands contaminate food and/or food surfaces.
- Food contaminated by dust, air and water containing spores, or faeces of insects, rodents or animals.

## Control measures
1. Good personal hygiene; hands must be washed; staff suffering from diarrhoea must not handle food.
2. Clean and sanitise the kitchen to eliminate dust, soiling etc. in which spores survive.
3. Keep contaminable items separate. Soil from potatoes, for example, should be kept out of the preparation areas. Storage areas should be kept clean.

4 Cook food (above 65°C) thoroughly and **rapidly** to kill the bacteria before spores are produced.
5 Cool food (below 5°C) thoroughly and **rapidly** to limit its time in the danger zone and restrict bacterial growth to a minimum. This may involve placing the food in small shallow containers to allow heat dissipation and cooling in the refrigerator.
6 Where possible cook food just prior to serving and eating. Beware of leftovers.

## A case history

A keen gardener assigned to a restaurant used powdered animal manure on the restaurant's gardens and lawn. A lawnmower carrying particles of this animal manure was placed in the corridor to the kitchen. Fine manure particles containing *Clostridium perfringens* spores were introduced by staff traffic and air movement etc, into the adjacent cold room, where fresh meats, stored almost at floor level, became contaminated.

The chef generally undercooked the large meat roast so that the centre (anaerobic conditions) still showed signs of blood. The spores vegetated and the bacteria produced were at an ideal temperature for rapid growth. The roast was placed in a smorgasbord display where the bain marie's heating element was defective and the bacteria in the meat continued to multiply rapidly all through that evening. At the end of the first night the roast was placed in the cold room up against other roasts (ham, lamb and pork). These became cross-contaminated.

The next night all the roasts were partly heated and placed in the defective bain marie in the smorgasbord display area, promoting further growth. On both these nights, patrons came down with food poisoning. On the third day the roast meat remaining was made into a meat pie. The remaining bone and any adhering meat were made into stock by slowly bringing to the boil. Spores were produced which survived the boiling process. The stock was allowed to cool at room temperature, which enabled spores to vegetate and grow rapidly. The contaminated stock was used to make soup, entrees, potato pie, and the already contaminated beef pie. This night most of the numerous food items on the smorgasbord were contaminated. The majority of the patrons were stricken with food poisoning.

Keeping unsuitable equipment out of the kitchen, thorough cleaning and sanitising of the kitchen, thorough cooking, maintaining the equipment in good working order and preventing cross-contamination of foods would have prevented these problems.

## 4 *Bacillus cereus* infection/intoxication

### Causative organism

*Bacillus cereus*

### Description

Rod-shaped bacilli

## Habitat
Soil, dust, cereal and flour, spices, vegetables

## Growth characteristics

*Temperature*
10–49°C (optimum 30°C)

*Oxygen*
facultative (aerobic or anaerobic)

*Moisture (Aw)*
0.93

*pH*
4.9 to 9.3 (optimum 7.0)

## Special characteristics
1. Infective dose is large. Sometimes up to 100 million micro-organisms/g food are required before they cause infection.
2. The bacteria produce spores resistant to boiling.
3. Toxin produced in two forms:
   (a) released in the food before consumption (vomiting form)
   (b) released in the intestine (diarrhoea form).

## Foods involved
Cereal products, rice, custard, cornflour and sauces, meatloaf. Problem exacerbated by bulk cooking of food.

## Mode of transmission
- Dust introduced into kitchen.
- The bacilli may already be present in food in spore form, just needing the right environment to stimulate their growth and toxin production.

## Control measures
1. Kitchen hygiene — keep premises free of dust etc.
2. Rapid cooking and rapid cooling of foods (similar to *Clostridium perfringens*).
3. Do not leave food at room temperature. Remember to keep food hot (above 60°C) or cold (below 5°C). This will not allow the bacterial spores time to vegetate and grow (in the danger zone) to the infective dose and then produce the toxin.
4. Where possible cook food just prior to serving and eating.

# A case history
A cafeteria prepared custard for a dessert. The dry custard powder (containing spores of *Bacillus cereus* when purchased) was added to a very large container of boiling milk. The spores withstand boiling

but vegetate and grow when the custard cools and enters the danger zone.

The custard was decanted into one large container and several smaller containers and left at room temperature (within the danger zone) for a period of time before being placed in the cold room. The small containers cooled quickly but the large container held the heat, even in the cold room.

The temperature of the custard in the large container remained in the danger zone until served several hours later. The bacteria grew to the infective dose then produced a toxin and food poisoning resulted. The smaller containers did not have sufficient growth to reach the infective dose.

This outbreak could have been prevented by decanting all the custard into smaller, shallow containers. This would enable the heat to dissipate quickly and the small containers could be refrigerated almost immediately to produce rapid cooling. Even though the bacteria were initially present in the dry custard powder, rapid cooling would not allow the vegetative bacteria sufficient time to grow to the infective dose.

## 5 *Vibrio parahaemolyticus* infection

### Causative organism
*Vibrio parahaemolyticus*

### Description
Marine vibrio

### Habitat
Coastal seawater and sediments in tropical, sub-tropical and temperate waters above 10°C.

### Growth characteristics

*Temperature*
5–43°C (optimum 37°C)

*Oxygen*
facultative

*Moisture (Aw)*
marine organism, capable of living in salt water (halophilic)

*pH*
5.0 to 9.6 (optimum 7.5 to 8.5)

### Special characteristics
1  Susceptible to heat. *Vibrio* destroyed above 65°C.
2  Susceptible to cold. *Vibrio* destroyed below 0°C.

3 Halophilic (loves salt). Requires salt for growth with an optimum concentration of 3%.
4 Infection occurs at an infective dose of 1 million micro-organisms/g of food.
5 No carrier state exists. However food handlers' hands may cross-contaminate foods.

### Foods involved
Prawns, crab, lobster, seafood cocktails and seafood eaten in cold state whether precooked (cross-contaminated) or not.

### Mode of transmission
- Insufficiently cooked seafood.
- Cross-contamination e.g. from raw to cooked seafood,
    from cutting boards or equipment to food,
    from food handlers' hands to food.
- Using seawater or brackish water to thaw frozen seafood.

### Control measures
1 Cook above 65°C to kill the *Vibrio*.
2 Prevent cross-contamination by using standard personal and food hygiene practices.
3 Use standard thawing procedures. Do not use seawater or brackish borewater to thaw frozen fish or shellfish.
4 Store seafood below 0°C to prevent growth to the infective dose if food is to be eaten raw, semicooked or precooked.

# A case history

To supplement the fresh water supply on an island resort, unsatisfactorily chlorinated brackish borewater was used for toilet and ablution facilities as well as for washing-up facilities in the kitchen of the resort.

Unknown to anyone, *Vibrio parahaemolyticus* was present in the brackish borewater.

It was the practice of the kitchen staff to thaw precooked frozen foods such as shelled prawns, crab and lobster meat by immersing the frozen blocks in a sink of borewater (with a salt concentration of 1.4%) and to leave the thawed meat floating in the borewater at room temperature (30–38°C) for an indeterminate time.

Under these circumstances, *Vibrio parahaemolyticus* contaminated the precooked seafood and conditions were ideal for their very rapid growth. The seafood was served cold, without undergoing further cooking, and a massive food-poisoning outbreak resulted.

This outbreak would not have occurred if recognised thawing procedures had been carried out, or if the seafood had undergone further cooking, which would have destroyed the *Vibrio*.

# Personal hygiene and the food handler

Personal hygiene has special meaning for food handlers. Normal **unconscious body habits** such as scratching an itchy nose, face, hair or body generally don't create any problems in normal life but can lead to catastrophe when handling foods.

## Personal health

Disease can be spread by carriers, that is, healthy people who have particular, harmful bacteria on or in their bodies but who do not suffer any harmful effects. This is the case with:

*Staph. aureus* — at least 70% of our population carry this micro-organism normally in the nose without symptoms.

*Salmonella/Clostridium perfringens* — can be present in the intestine of apparently healthy people.

People suffering from

- a contagious disease
- an upper respiratory problem, for example sinus or a cold
- an intestinal upset, for example diarrhoea

shed the bacteria at a greater than normal rate. The bacteria use body fluids and water as a means of transport.

One of the major problems with food service managers is that they often insist on staff working when they are genuinely sick to 'get their pound of flesh', putting their complete business in jeopardy by creating a high risk of food poisoning.

> If you are sick, do not handle food.
> Report it to your manager. Stay home until you're better.

◀ 34  Food Poisoning Prevention

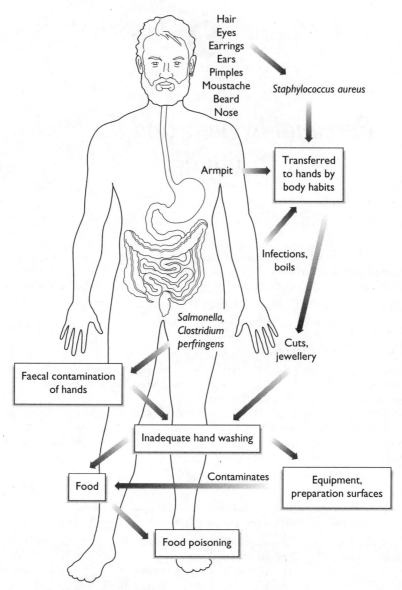

**Figure 7.1**  The food handler as a carrier

We described food-poisoning bacteria in chapter 3 as a 'time bomb'. Humans are 'walking time bomb carriers' occasionally setting them off.

Everyone has to assume they may be a **carrier** and guard against spreading disease by being clean in body, habits and clothing. For bacteria to spread, they need a way and a means to 'hitchhike' from our body to food. The way is by bad body habits and the means is generally by our hands.

## *Body habits*

Avoid the following actions while handling food.

1. Scratching any part of the body or clothing or combing the hair.
2. Wiping perspiration from the body, e.g. rubbing the armpits.
3. Unconsciously stroking moustaches or beards or the edges of the mouth during conversation. This spreads *Staph. aureus* onto the hands.
4. Unconsciously rolling the sleeper in pierced ears during conversation, spreading *Staph. aureus* onto the hands.
5. Smoking in the kitchen area. Saliva from the mouth is transported from the cigarette to the fingers and from the fingers to the food. This is equivalent to spitting directly onto the food. Handlers smoking outside the kitchen should wash their hands before resuming work.
6. Coughing or sneezing directly onto food or into the hand. Use tissues but still wash the hands immediately, as bacteria pass through tissues.
7. Rubbing the index finger along a 'running' nose. Use a tissue. Anyone with a heavy cold should stay at home or keep away from food preparation areas.
8. Picking pimples. These are a rich source of *Staph. aureus*, which get caught under the fingernails.
9. Tasting food by using the fingers or a ladle that is returned to the food with saliva on it. Use a new spoon each time or ladle some food into a tasting dish.
10. Licking the thumb or fingers to separate wrapping papers. This transfers saliva onto the paper or the next lot of food handled.

## *Keeping hands clean*

Hands are by far the most common mode of contamination.
Hands and forearms should:

1. be free from minor cuts or abrasions. Regularly changed waterproof dressings or disposable gloves may assist if cuts or abrasions are present.
2. be free from infections, boils or dermatitis. Stop work until the infection has gone.
3. have nails well manicured, preferably trimmed short and free from nail polish. Nail polish may chip into food and may also cover up filthy nails.
4. be free from jewellery as food and bacteria get caught in them causing cross-contamination. Stones may also fall into the food.
5. be washed thoroughly by
   - pre-rinsing to remove soiling
   - washing in a rich lather preferably twice, using liquid soap and warm water

- brushing under the nails, but not so vigorously as to cause abrasions. Nail brushes can contaminate if not cleaned and sanitised regularly.
- rinsing then drying with disposable hand towels or a hot air dryer. Tea towels or towelling hand towels used by everyone may be a reservoir of contamination.

## When hands should be washed
- before and after commencing work.
- after visiting the toilet. (*Salmonella* and *Clostridium perfringens* in the faeces may get through toilet paper and be transferred to food via the hands.)
- after unconscious body habits (see earlier).
- after blowing the nose. (*Staph. aureus* may get through handkerchiefs or tissues and be transferred to food via the hands.) Dispose of tissues sensibly after use.
- after handling refuse or contaminated foods.
- before and after handling raw produce such as meat, poultry and vegetables, as bacteria or spores may be transferred to other foods, for example *Clostridium perfringens* spores on vegetables.
- before and after handling different types of food, for example uncooked and cooked (to prevent cross-contamination with *Salmonella, Clostridium perfringens, Staph. aureus* and *Vibrio parahaemolyticus*).
- after handling money. Contrary to belief, coins and notes do not greatly support bacterial growth; however they can easily transport viruses, or bacteria requiring small infective doses.
- after handling returned soiled glasses, crockery and cutlery. These items have come in contact with the customers and may contaminate food handlers. It is a common practice for waitresses to carry several soiled glasses by the rim and in so doing contaminate their hands with the saliva from the customers' mouths.

# 8

# *H*azard analysis of food

This chapter contains a broad list of foods, the most likely bacteria found in these foods if contaminated and the possible sources of contamination.

> - Always assume food received from a supplier may already be contaminated.
> - Don't assist bacteria to grow and reach the infective dose by mishandling.

## *P*otentially hazardous foods

Evaluate the risk of food poisoning in the food you prepare. List all foods with their ingredients and the method of preparation used. Using Table 8.1, identify the bacteria most likely to be present. These will usually be *Salmonella, Staphylococcus aureus, Clostridium perfringens, Bacillus cereus, Vibrio parahaemolyticus*. Now check their individual growth and special characteristics as well as the control methods given for each in chapter 6.

Know your enemy. Ensure that the correct control measures are taken for each particular food. Sometimes a change in the way you prepare a food may eliminate the problem. Remember, the *B. cereus* case history. Just by replacing the larger pot with smaller shallow containers to cool the custard would have prevented the food-poisoning outbreak.

## Evaluating beef stew
This is given as an example (refer to Table 8.1).

### Ingredients
Meat — *Salmonella, C. perfringens, Staph. aureus*
Various vegetables — spores of *C. perfringens*
Salt — inhibitor of bacterial growth
Seasonings/spices — spores of *C. perfringens*
Water — potable (nil)

## Method
Cut up, bring ingredients to boil slowly, simmer for one and a half to two hours, cool on bench till cold, place in cold room in original container. Sufficient for two days' supply.

## Evaluation
If any ingredients are contaminated, the most likely culprits are *Staph. aureus* and *C. perfringens*.

Look up the characteristics of both (see chapter 6, B and C).

*Salmonella* — boiling kills *Salmonella* and eliminates it completely.

*C. perfringens* — with slow boiling, *C. perfringens* in meat may multiply and have time to spore before reaching high temperatures. Spores may also be present in spices and partly washed vegetables, and can withstand four hours of boiling. Leaving food to cool at room temperature allows spores to germinate back to the vegetative state when the danger zone is reached and then multiply rapidly. Placing the stew in the large container (even in the cold room) means the stew is kept in the danger zone and ensures extremely high growth in number of bacteria, sufficient to cause food poisoning.

*Staph. aureus* — boiling will kill the bacteria. However if bacteria have had sufficient time to produce toxins, boiling will not destroy these toxins.

Risk of food poisoning — high.

## Correct method
Work quickly at room temperature. Thoroughly wash all vegetables to remove soiling, cut up all ingredients, rapidly bring to boil, then simmer for one and a half to two hours. Cool slightly, then rapidly cool by putting stew into smaller shallow containers. If stew is to provide meals for two days, divide the quantity made into multiples of two. Refrigerate immediately in cold room. Don't top up the second day's supply with the first day's leftovers.

It is however preferable to make up sufficient for one day only and keep the stew hot until served. This prevents the spores from returning to the vegetative state.

**Table 8.1** Foods that support bacterial growth

| Food | Most common bacteria | Contamination/Survival |
|---|---|---|
| *Meat and poultry* | | |
| Fresh meat | *Salmonella* | Contaminated during evisceration |
| | *C. perfringens* | Present in large cuts of meat |
| | *Staph. aureus* | Contaminated by food handler |
| Mince | As for fresh meat | Bacteria now evenly distributed through mince |
| Rolled roasts | *C. perfringens* | Spores survive slow or under-cooking |
| | *Salmonella* and *Staph. aureus* | Contaminated surface rolled into centre of roast. Survive if under-cooked |
| Reheated foods | *C. perfringens* | Slow cooling and/or reheating; reheating temperature not high enough |
| | *Salmonella* | Under-cooked when reheated |
| Cold meats | *Staph. aureus* | Contamination by food handler |
| | *Salmonella* | Cross-contamination — meat not reheated |
| Cured meats | *Staph. aureus* | Survive high salt content when cured |
| Fresh poultry | *Salmonella* and *C. perfringens* | Mainly in chest cavity |
| | *Staph. aureus* | Contaminated by food handler |
| Roast chicken | *Salmonella* | If frozen, not fully thawed; under-cooked in thicker flesh areas |
| Boiled chicken | *C. perfringens* | Spores survive boiling |
| Cold leftover chicken | *Salmonella* | Cross-contamination |
| | *Staph. aureus* | Contamination by food handler |
| Stocks | *C. perfringens* | Spores survive boiling |
| Stews/Casseroles | *C. perfringens* | Spores survive boiling |

| Food | Most common bacteria | Contamination/Survival |
|---|---|---|
| ## Milk and dairy products | | |
| Raw milk | Staph. aureus<br>Salmonella | Mastitis in cows<br>Contamination by manure |
| Pasteurised milk | No contamination | Subject to contamination after opening |
| Powdered milk | Salmonella and Staph. aureus | Survive drying process |
| Pasteurised cream | Staph. aureus | Contaminated after opening |
| Raw cream | Staph. aureus and Salmonella | As for raw milk |
| Cheese | Staph. aureus | Contamination by cheese cutters |
| Butter | Staph. aureus | Contaminated after opening |
| ## Egg and egg products | | |
| Hen eggs | Salmonella | Faecal contamination of cracked eggs |
| Duck eggs | Salmonella | Contamination from infected ovaries; faecal contamination of cracked eggs or eggs laid in moist conditions |
| Dried, liquid or frozen eggs | Salmonella | Contaminated eggs, if not pasteurised |
| ## Fish and seafood | | |
| Fresh fish | Salmonella and C. perfringens<br>C. botulinum (E)<br>V. parahaemolyticus | Filleted and cleaned in waters polluted by sewerage outfalls<br>Overseas outbreaks<br>Present in seawater |
| Smoked fish | C. botulinum (E) | Mild smoking — overseas outbreaks |
| Raw fish | V. parahaemolyticus | Present in seawater — common in Japan |

| Food | Most common bacteria | Contamination/Survival |
|---|---|---|
| Fish rissoles | *Staph. aureus* | Contaminated by food handler when moulding |
| Raw oysters | *Salmonella* and *Escherichia coli* | Use of contaminated 'flushing' water during cleaning |
| | *V. para-haemolyticus* | Present in seawater |
| Prawns, crabs, lobster, Moreton Bay bugs | *V. para-haemolyticus* | Cross-contaminated after cooling |
| | *Staph. aureus* | Contaminated by handlers during shelling |

## Canned food

| | | |
|---|---|---|
| Canned fish, seafood and vegetables; home-bottled vegetables | *C. botulinum* (E) | Defective canning; cans subjected to minimum temperatures to retain flavour and appearance. Generally overseas |
| Opened canned food | *Salmonella*, *C. perfringens* and *Staph. aureus* | Contaminated by handler. Contaminated if not refrigerated immediately |

## Pasta and rice

| | | |
|---|---|---|
| Boiled/Fried | *B. cereus* | Spores survive slow cooking and cooling of rice and pasta |

## Vegetables

| | | |
|---|---|---|
| Wide variety of vegetables especially those grown in animal manure | *C. perfringens* | Contamination by soil; vegetables act as 'vehicle' only, contaminating other foods, e.g. cabbage leaves used in food displays to surround cold meats etc. Spores survive boiling |

## Desserts and sweets

| | | |
|---|---|---|
| Ice cream | *Salmonella*, *C. perfringens* and *Staph. aureus* | Contamination after processing by handlers; thawing and refreezing aggravates problem |

| Food | Most common bacteria | Contamination/Survival |
|---|---|---|
| Imitation cream | *Salmonella* | As for powdered milk |
|  | *Staph. aureus* | Treat as for cream |
| Custard | *B. cereus* | *B. cereus* spores survive boiling |
|  | *Staph. aureus* | Contaminated by handler |
| Trifles | *B. cereus* | From custard powder |
|  | *Staph. aureus* and *Salmonella* | From cream |
| Coconut | *Salmonella* | Processing defects; survive light cooking temperatures, e.g. macaroons |
|  | *C. perfringens* | Spores survive desiccation |

## *Food additives, sauces and dressings*

| Food | Most common bacteria | Contamination/Survival |
|---|---|---|
| Spices | *C. perfringens* | Spores survive cooking |
| Gelatine | *Salmonella* | Survive light cooking or glazing after cooking |
| Stocks | *C. perfringens* | Spores survive cooking |
| Gravies | *C. perfringens* | Spores from cooking meat etc. survive in gravies |
|  | *B. cereus* | Spores survive in flour (thickener) |
| White sauces | *B. cereus* | Spores survive in flour (thickener) |
| Mayonnaise and dressings | *Salmonella* | Survive light cooking or in ingredients added after cooking. No cooking, e.g. raw eggs used in Hollandaise sauce |
|  | *Staph. aureus* | Contaminated by food handler |

**Table 8.2** Foods less likely to support bacterial growth or cause food poisoning

Dried foods — with a reduced moisture content (Aw)

Jams, honey, confectionery — with high sugar content

Fats and oils — with high concentrations of fats

Sauces, pickles and chutneys — owing to the low pH of these acid-type foods

Alcohol — depends on the type and quantity of alcohol used in the cooking process

Bread — crust helps protect from contamination

Note: These foods may, however, support mould growth.

# An introduction to HACCP

HACCP is an acronym for Hazardous Analysis, Critical Control Points. HACCP was pioneered in the early 1960s for the United States NASA Space Program and is now recognised as the safest, most effective food safety system in the world. Consequently, food safety legislation is now incorporating HACCP on an international level.

HACCP systems are designed to be pro-active in preventing the incidence of possible food safety problems within the food industry by analysing the inherent risks to a product or process and then deciding what control measures can be put in place to overcome these identified risks. The hazards include biological, chemical, or physical hazards. Biological hazards are the major cause of foodborne illness. In analysing these hazards, 'hot spots' (**critical control points**) within a process or recipe are identified.

Once food managers become aware of the likely **hazards**, it becomes more obvious what **control measures** should be in place, what needs to be **monitored**, and within what **limits** as well as what **corrective action** should be taken when the control measures are not met. Often where uncontrolled hazards are identified within a process, adjustments can be made to correct them without additional cost to management.

Effectively food safety education and training is the key to successful HACCP implementation. Unless employees are well aware of why they are carrying out a procedure and which steps are critical, a hazard may loom.

## The seven principles of HACCP

The American National Advisory Committee on Microbial Criteria for Food (NACMCF) and Codex Committee on Food Hygiene developed seven principles for the implementation of HACCP by the food industry. It is on these principles the modern HACCP food safety plan is based.

### Principle 1   Hazard analysis

The hazard analysis process identifies significant hazards likely to occur and preventative control measures needed to be implemented to overcome these hazards.

The standard approach is to assemble a team that has a knowledge of the food processes and understand food safety principles. This team identifies the biological, chemical or physical hazards associated with each and every food handled or prepared that have the potential to cause unsafe food.

A flow diagram sets out the various steps of the process from the acceptance of food into an establishment through to the release of the food to the customer. The flow diagram gives a better visualisation of the process.

Bacterial pathogens comprise the majority of foodborne outbreaks throughout the world. By targeting biological hazards and developing effective procedures the greatest impact on food safety is achieved.

Chemical hazards should be considered where naturally occurring chemicals or toxins are likely to exist in the food or where chemicals find their way into the food during the growing, harvesting or processing.

Physical hazards are more noticeable and physically screened out, for example a stone in flour.

The determination of risks is based on the team's knowledge of food safety, taking into account scientific evaluations. Risks are associated with:

- sensitive ingredients (potentially hazardous food or chemical/physical hazards finding their way into the food)
- possible microbial load of the food
- intended use and intended consumer's risk group (for example food for an aged care home is high risk)
- staff's health hygiene practices and food safety knowledge
- intrinsic factors of the food
- conditions of storage
- preparation/processing procedures
- packing or serving of the food
- facility design
- equipment design
- cleaning and sanitising procedures.

# Principle 2   Identifying the Critical Control Points within the process

Critical Control Point (CCP) is defined as **any point or procedure in a specific food system where loss of control may result in an unacceptable health risk.**

Many points in food preparation may be considered control points but very few are actually **critical** control points. The critical control point decision tree is shown in appendix 4.

Different premises for preparing the same food may have different procedures and consequently different critical control points. Therefore generic HACCP plans can only be used as a possible guide as it is the unique conditions with each facility or process that should be considered in the development of a HACCP plan. It is for this reason that I have developed the *HACCP Master* food safety computer program.

## Principle 3   Establishing the critical limits for preventative measures

Setting the **critical** limits for monitoring control measures means taking into consideration the minimum and maximum values at the critical control point to minimise the identified risk. Historical scientific data may be available on desirable microbiological critical limits. If not, microbiological testing may be carried out initially to identify microbiological safety limits in the procedure or process. Chemical limits are usually set by food standard codes and physical limits should be zero or nondetectable.

## Principle 4   Establishing procedures to monitor critical control points

Monitoring is a planned sequence of either observation or measurement of some form to assess whether the critical control points are under control. Monitoring will indicate whether there has been a loss of control and checking of historical monitoring information. A decision can be made about what corrective action to take, whether it be reprocessing, treating or discarding the food.

Continuous monitoring of critical control points will always be the best option, however this is not always practical. Calibration of monitoring equipment is vital. Management nominates **responsible persons** to monitor and record the information so that immediate corrective action can be taken where the process exceeds the critical limits.

Monitoring is a crucial step towards self-assessment and self-management. Self-assessment/audit has been the missing link as well as past low performance of the old food inspection system where complete reliance on food inspections by an external agency gave management no control over food safety problems.

## Principle 5   Establishing the corrective action to be taken

Where monitoring indicates that the critical limit has been exceeded, corrective action would need to be taken by the responsible person. This would take into consideration:

- the status of any food when the problem occurred
- what is required to be carried out to resume control or abort the process batch
- how to overcome the cause of the problem to reduce future incidence where possible

- maintenance of records detailing corrective action taken, by whom and its effectiveness.

## Principle 6   Establishing and maintaining records to document HACCP

Part of the record keeping would include a written master HACCP food safety plan that would identify the critical control points, responsible personnel and supervisory/verification personnel, intended use and a flow diagram for each food prepared. At every CCP there should be listed the hazards, the control measures along with critical limits, the monitoring and verification procedures and corrective action undertaken. Records generated during the operation are required to be logged to assist in the auditing and verification process.

## Principle 7   Verification of HACCP

The verification process indicates whether or not procedures are being carried out in accordance with the HACCP plan and whether it is working effectively. Where modifications are necessary, records of who authorised the modifications and the reason for the modifications are required.

The verification process may be carried out by the Regulatory Authority with an audit of records and an inspection to ensure the HACCP food safety plan is adhered to. Sometimes random sampling assists.

Part of the verification process is a review of education and training of staff. Food safety knowledge plays a key role in the implementation of HACCP and training should be carried out to ensure that staff understand and apply the principles of food safety. Unless management ensures responsible persons have adequate knowledge and training to carry out the duties they are responsible for, the HACCP plan will be useless. A HACCP food safety plan is a 'live', working document.

## *The practical implementation of HACCP*

**The most difficult and time consuming process under HACCP is identifying critical control points.** The standard team approach is almost impractical in small business, especially when faced with the additional fact that the majority of the staff may not be adequately educated in food safety and do not recognise the likely hazards.

Within seconds, the *HACCP Master* computer program **automatically evaluates food recipes and processes under HACCP principles to identify control points (CP) and critical control points (CCP)** within each and every food product that is handled or prepared. Processes can be edited and re-evaluated as often as required, to ensure the process is technically correct. At each identified control point or critical control point, possible hazards are identified to enable the implementation of control measures, desirable limits, monitoring and verification procedures as well as corrective action where control measures are not adhered to.

*HACCP Master* is an inexpensive but powerful tool for the food industry to confidently create, own and implement their food safety program especially where numerous food products are prepared and regularly changed to suit clientele.

To assist food managers in creating a cleaning and sanitising program under their HACCP food safety plan, I have developed *Cleaning Master*, a cleaning and sanitising scheduling computer program that allows management to compile a cleaning and sanitising master schedule for their premises, allocating detailed procedures to responsible persons and supervisors at nominated times.

*HACCP Master* and *Cleaning Master* can be purchased from Envirohealth, PO Box 6093, Rockhampton Mail Centre (4702) Queensland, Australia. Phone (+61) 079 28 4657; Fax (+61) 079 28 8135.

# Food hygiene and food handling

Food hygiene is the manner in which food is handled or managed, so that food poisoning is prevented.

This is done by:

- protecting the food from contamination
- either preventing bacterial growth from reaching the infective dose, or destroying the bacteria if, in each case, the food has previously been contaminated.

We shall now consider ways of achieving this by looking at the following:

- the purchasing of food supplies
- the storage of dried foods, canned foods, vegetables, refrigerated food and frozen food
- the preparation of food
- cooking processes
- the display and serving of food
- storage and reheating of leftover food.

## Purchasing food supplies

1 When purchasing food, deal only with reputable suppliers thus ensuring quality control from a suitably approved source that uses correct handling and storage procedures.
2 Use the senses of sight and smell when inspecting incoming food. Tasting is not recommended. The product may be contaminated. Check for:
    - undesirable change of appearance or smell, for example mould on bread or rancid milk smell
    - physical damage to packaging
    - signs of thawing or incorrect delivery temperatures, for example frozen food should be received in a frozen state
    - evidence of insect or rodent interference
    - any variation from the normal in colour, texture, odour or general appearance that may warrant further investigation.

Refuse to accept delivery should a problem arise and if in doubt request further advice from the delivery person, supplier, or local environmental health officer.
3. Arrange delivery times so you can personally inspect, accept and store items. Don't allow food, for example bread and milk, to be left outside the door where it can be interfered with by passersby, or contaminated with dog or cat urine, dust or other environmental elements.
4 Delivery personnel have to observe the same personal hygiene and handling procedures you are required to uphold. If they don't, draw it to the attention of your supplier, or local environmental health officer and change your supplier if not satisfied. Food poisoning can ruin not only their business but yours as well.

## *S*torage of food

Compliance with the 'use by' date and rotation of stock using FIFO (first in, first out) are necessary to ensure premium quality.

## Dried food

This should remain sealed and be kept in a cool, dry, well ventilated environment so it does not become partly reconstituted and subject to bacterial growth. Inspect to ensure it is free from contaminants and insects that infest stored food (chapter 12).

Once reconstituted, treat as a potentially hazardous food, and refrigerate (below 5°C).

## Canned food

Canned food should not be

- in misshapen, swollen or 'blown' cans, which may indicate bacterial or chemical action and the production of gas.
- in cans which are leaking, dented, rusted or contain poor structural seams. Fractures in the score lines of lids of 'pulltop' containers allow the vacuum seal to be broken, and air and/or bacteria to enter.
- allowed to remain at room temperature after opening. Treat as potentially hazardous food. Leftovers should be placed in a sanitised container and refrigerated (below 5°C).

Some canned foods (for example mushrooms) undergo only minimal heat treatment, so that the premium texture, flavour and appearance of the food are retained. Some canned foods may need to be refrigerated below 5°C, for example hams.

Read all labels, and follow instructions carefully. Under no circumstances buy a product without its original label.

Canned and bottled goods have a general shelf life of approximately one year when stored correctly. The same foods, when opened and refrigerated below 5°C, may only have a shelf life of approximately 2–3 days.

## Vegetables

Vegetables can be stored at room temperature, for example potatoes, or in a cool room (5 to 10°C), for example lettuce.

Some vegetables generally contain particles of soil and/or fertiliser fragments (animal manure) which can introduce bacteria into the kitchen. These foods should be thoroughly washed and their preparation should be kept separate, if possible, from the preparation of potentially hazardous foods.

## Refrigerated food

This type of food should be stored in a commercial refrigerator or cold room capable of maintaining the food below 5°C even under adverse conditions, such as frequent opening of the door in high room temperatures.

Never store cooked food under raw food. Drips or fragments of raw food falling onto the cooked food that may not be heat treated further, can spread contamination. Similarly, cooked foods should not come in contact with uncooked foods and transfer bacteria (cross-contamination). Do not stack uncovered food containers on top of one another thus allowing the bottom of the top container to contaminate the food below.

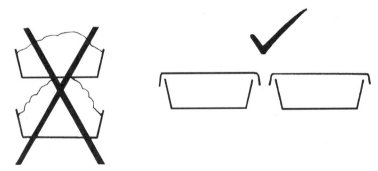

**Figure 10.1** Storing food in uncovered containers

Do not place food directly on shelves but in containers or trays that are sealed, covered or wrapped with flexible plastic wrap. This not only prevents contamination but reduces the absorption of strong odours. Foods with strong odours, for example seafood and some cheeses, should be sealed and separated from others that absorb odours quickly, for example bread and bakery products, milk, cream and dairy products.

Overcrowding a cold room or refrigerator prevents the air from circulating and the food may remain in the danger zone for a period of time. Keep food on shelving at least 30 cm above the floor to make cleaning easy, for example of food and milk spillages.

It is advisable to keep a probe thermometer available to check air and food temperatures at various positions in your cold room or refrigerator. Clean and sanitise the probe, using hot water or analytical alcohol after every insertion into the food, to prevent cross-contamination.

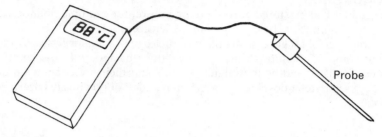

**Figure 10.2** A probe thermometer

When cooked foods are to be refrigerated or frozen, precool at room temperature for 15 to 20 minutes before placing in the refrigerator or freezer. Where large quantities or volumes are involved, package into batches or smaller volumes with large surface areas to dissipate their heat quickly before covering and refrigerating or freezing.

*Shelf life*
- Milk                           approx. 5–7 days
- Fresh meat                     approx. 3 days
- Poultry, mince and seafood     approx. 2 days
- Cooked or leftover food        1 day only, when stored between 0°C and 5°C.

## Frozen food

This should be kept below –18°C for long-term storage.

The problem associated with frozen food is that it may be thawed then refrozen, which not only shortens shelf life but also causes deterioration and possible conditions conducive to bacterial growth. In Australia it is illegal to sell ice cream that has been thawed and refrozen.

### Correct methods of thawing food

*In a refrigerator or cold room (below 5°C)*
This is by far the safest method. However, it requires both a reasonable amount of refrigeration space and forward planning (time to raise the temperature from –18°C to 5°C). The food is out of the danger zone for the entire period.

*In a microwave oven*
Select the proper defrosting cycle and time so that the temperature can be raised quickly from –18°C to approximately 5°C. Once thawed, the

food should be used immediately or placed under refrigeration until ready for use.

*As part of the cooking process*
Some frozen foods may be cooked directly in the frozen state. This is indicated in the cooking instructions on the package, for example of frozen vegetables and frozen dinners. Small cuts of meat etc. can also be cooked in this way provided they are cooked thoroughly by increasing the cooking time. The food has to pass quickly through the danger zone.

## Risky methods of thawing
The period of time spent at a temperature in the danger zone can encourage rapid bacterial growth and cannot be recommended.

*Under running potable water (below 20°C)*
The food item should remain in the original waterproof and undamaged wrapper. There must be sufficient water velocity and the food should be placed on a wire rack; it should not be put directly in the sink nor immersed in water in the sink. The sink can be a haven for bacteria especially if it is also used for washing vegetables etc. or as a handbasin by careless food handlers.

*At room temperature (Have you checked your range of seasonal minimum and maximum temperatures?)*
Food items thawing at room temperature should be transferred to a refrigerator or cold room to finish defrosting when the outside surface reaches 5°C. However, at that stage the centres are still frozen. Food completely thawed at room temperature is generally left out some considerable time, becoming a potential 'time bomb'. This practice must be prohibited.

Any food, especially large cuts of meat or large poultry, not fully thawed require thorough cooking so that the centre of the thickest section of the meat or bird reaches 74°C, otherwise the centre is still in the danger zone, which encourages rapid bacterial growth. Poultry should be fully thawed, otherwise during the cooking process the chest cavity remains in the danger zone and stuffing acts like an insulation, causing ideal conditions for rapid growth of bacteria but not sufficient heat to kill them. These conditions suit *Salmonella* organisms.

Once thawed:
- food should be refrigerated and processed within a maximum of two days (preferably one day)
- cooked food should not be refrozen and, if reheated, should be eaten immediately or thrown out
- if thawed at room temperature food should be processed and eaten immediately or thrown out.

Points to look for to check whether frozen food has been thawed and refrozen:
- the presence of frosting on food

- the presence of ice slivers on packages; or food packages stuck together due to the refreezing of the condensation
- frozen peas, beans etc. should 'roll free' and not form in ice clumps
- refrozen ice cream has a gritty taste and may have ice slivers in it as well
- containers may be deformed when food is refrozen (for example ice cream containers). With thawed food, pressure exerted by stack packing causes the sides of the container to bulge and the lid to depress, spilling the food. Refreezing leaves the container in a deformed state.

## *Preparation of food*

A high standard of personal hygiene should be observed. Avoid **unnecessary human contact** with the food by using tongs, forks or disposable gloves as a **barrier** between the food and the food handler.

Strong perfumes on the wrist may be absorbed into food, for example when making sandwiches, and should be avoided.

Clean and sanitise as you go rather than leaving for the end of the day, as bacteria grow continually on food particles at room temperature and can be transferred onto other food being prepared (cross-contamination).

Do not use the same equipment or surfaces, for example knives and cutting boards, for both raw produce and cooked produce or even for different food items, as cross-contamination will occur.

Work quickly at room temperature, returning perishable ingredients to the refrigerator immediately they have been used.

> **If you are not sure of the quality of the food remember:**
> **When in doubt — throw out.**

Other points to remember:
- don't allow pets access to the kitchen
- don't put the tasting ladle back into the food
- don't smoke in the kitchen
- be conscious of body habits during preparation of food
- don't sit or lie on benches
- don't wipe your hands on your clothing, tea towel or absorbent hand towel but go through proper handwashing procedures.

## *The cooking process*

Food should be thoroughly cooked at temperatures between 74°C and 88°C.

Slow cooking and 'rare' cooking encourage rapid bacterial growth, especially when the food is not to be eaten immediately. Once food is cooked it should preferably be eaten immediately; if not, then keep it hot in a bain marie above 60°C or precool it, then refrigerate and keep cold below 5°C until required.

Bain maries and pie warmers only maintain temperatures and should not be used for heating lukewarm, cold or frozen foods. Before being placed in this equipment, food should be preheated to between 74°C and 88°C.

## *Displaying of food*

Food should be stored correctly and protected from contamination by the public.

Cold display equipment should be *prechilled* to below 5°C before transferring food to it from the cold room.

Hot display equipment should be *preheated* above 60°C before transferring cooked food to it.

For smorgasbords or self-service food, display equipment should be fitted with suitably designed sneezeguards to prevent droplet spread from coughing and sneezing by the public.

An attendant should also be available to supervise the public's handling practices (which can leave a lot to be desired); tongs etc. should be available for handling all food.

In self-service establishments, cold meats should be plated out and covered in transparent plastic wrap to prevent contamination.

The use of individual prepacked items such as butter, jams, honey etc. also prevents contamination.

Ensure decorative displays using vegetables, flowers etc., especially in cold meat displays, do not touch or contaminate foods.

## *Serving of food*

The servery staff are the front line to the public and represent the hygiene of any premises. If they are immaculate in dress and use hygienic food-handling practices, customers assume the premises are well maintained. The opposite also applies.

### *Hygienic food-handling practices*
- use tongs, spatula, slice, spoon or scoop when handling food as a barrier between the hands and the food being prepared. Don't use hands.
- use disposable gloves etc. when handling potentially hazardous foods in the cold state.
- when serving food never handle parts of the cutlery or crockery that will come into contact with food or the customer's mouth. This *protects the customer from you.*

   Hold: – plates by the base
   – cups by handle or put on a saucer
   – glasses by the base or put on a tray
   – cutlery by the handles.

In retrieving soiled dishes etc. the same procedures should be practised *to protect you against the customer.*

- wash hands after cleaning tables, after handling money or before serving
- inspect glasses, crockery and cutlery for soiling, chips or other defects
- if wrapping food, don't lick your thumb before separating sheets of paper
- ensure drinking straws are protected in an approved dispenser
- invert takeaway food containers until filled to prevent dust or foreign objects dropping into them (Figure 10.3).

**Figure 10.3**

## *Leftovers*

**Eating leftover foods can be like playing Russian roulette.** It is most likely that such foods have been subjected to contamination and it only needs sufficient time for the bacteria to reach the infective dose.

It is imperative that leftovers are cooled quickly; if they are to be reheated, they should be reheated to between 74°C and 88°C rapidly and thoroughly.

Leftover food such as soups and stews should not be used to top up tomorrow's fresh batch. Throw it out. Custards and cream-based foods should not be used as leftovers.

Leftover food must be utilised within 24 hours or thrown out. If leftover food is to be used for stock, the stock should be made immediately.

# Decontamination — cleaning and sanitising

## Some definitions

### Cleaning
The removal of unwanted visible material (such as grease, food, dust, stains and other contamination), and of odours and flavours.

### Sanitising
The destruction of pathogenic (disease-causing) micro-organisms or the reduction of their numbers to a safe level. Sanitising does not destroy spores.

### Sterilising
The total destruction of all micro-organisms including spores. This highest standard of decontamination is rarely required in a food establishment.

By analysing bacterial swabs it has sometimes been found that kitchen floors and toilets are cleaner and safer than the preparation benches.

Food handlers consider kitchen floors and toilets as 'germy' areas and put more effort into cleaning and sanitising them than the preparation areas. If preparation bench tops are merely cleaned, bacteria survive and/or grow on food particles or juices on them.

Wiping a surface merely *evenly distributes* the bacteria over it.

## Principles involved

These include precleaning, cleaning, sanitising and drying, and apply equally to:

- premises with floor areas that are moppable only or fully hoseable
- fixtures and fittings such as preparation benches
- food equipment such as a meat slicer
- crockery, cutlery etc.

Carry out the following simple procedures and you will be able to guarantee your customers safer food.

## Precleaning (preparation stage)

1. Clean the area or dismantle equipment for thorough cleaning.
2. Lightly remove food scraps by scraping food off plates to remove bulk contamination. Store scraps correctly in the garbage area.
3. Pre-rinse in cold water to further remove food particles. This prevents the food particles from reducing the detergent's cleaning power during the later cleaning process. Cold water should be used for foods such as egg and blood products which are 'baked' onto the surface if hot water is used.

Warm water may be used to pre-rinse greasy surfaces.

## Cleaning (washing stage)

Using a suitable detergent or detergent/sanitiser and hot water (at least 50°C), wash all items. This cleans and removes grease and food traces, but not all the bacteria.

Effectiveness of cleaning depends on:

- the degree of soiling remaining after precleaning and the type of surface material to be cleaned
- the suitability of the detergent used
- the contact time
- the pressure applied.

## Sanitising (rinsing stage)

You cannot effectively sanitise a dirty surface that has not been adequately cleaned previously, as the soiled surface will deactivate the sanitiser before it can kill the bacteria. Very hot water (around 82°C) is generally the cheapest way to sanitise a surface. This is not only a combined sanitising and detergent rinsing process but also aids drying. Check the food hygiene regulations for the minimum **temperature** and **contact time** required. The higher the temperature of the rinsing water the shorter the contact time required.

If a chemical sanitiser is used another three stages are involved:

- rinsing between washing and using the sanitiser to remove traces of detergent
- using sanitiser at the **correct strength** and **correct temperature** and for the **desired contact time** to kill the bacteria
- rinsing preferably with very hot water to remove traces of sanitiser. The heat of the water aids the drying process.

The chemical sanitisers generally used are chlorine, QAC (quaternary ammonium compounds) and iodine. Refer to the detailed instructions for use on the label of the container.

Alcohol (methylated spirits) is another sanitiser that is easy to use and does not require further rinsing (owing to evaporation). A 70% methylated spirits spray is excellent for preparation benches and some equipment where liberal amounts of rinsing water cannot be used.

## Drying and storage

If sufficient heat is absorbed during the sanitising process, the item will air dry. Using tea towels, unless they are clean, may only recontaminate the already sanitised surface.

When chemical sanitising requires a cold rinse, for example in washing down benches or floors in very large kitchens, the surface may need to be dried using a previously sanitised squeeze.

Once a surface has been sanitised and dried, protect the item as far as practicable from further contamination, for example by correct handling or by storing crockery and cutlery in vermin-proof storage cupboards.

The most contaminable items in the kitchen are cleaning equipment such as dirty dishcloths and cleaning cloths, scrubbing brushes, brooms and mops. These items may allow food to cling to them during the pre-cleaning/cleaning process and if they are not cleaned and sanitised correctly, bacteria survive and multiply in the food on the cloth or mop. The next time these are used, the bacteria are spread over the surface being cleaned.

Cleaning equipment should not be left around a kitchen but placed in a special cupboard or cleaner's room.

## Cleaning and sanitising program

All food premises, no matter how big or small, require continual cleaning and sanitising to eliminate food-poisoning bacteria before they become a 'time bomb'. Certain areas, equipment and surfaces are more prone to contamination than others and may require more frequent attention or specialised procedures.

All premises need a program to ensure that they are completely and adequately cleaned and sanitised.

Do you have a complete cleaning and sanitising program for your premises? If not, then this is how you can devise one. The *Cleaning Master* computer program by Envirohealth (Ph: 079 28 4657) is an excellent tool for managers to compile a cleaning and sanitising master schedule for their premises. It allocates detailed procedures and requirements for those in charge at nominated times and gives the flexibility to cope with changing staff or responsibilities.

**Table 11.1:** Master cleaning and sanitising program

| Item | Procedure | Materials/Products | When | Person responsible |
|---|---|---|---|---|
| Kitchen floors | 1 Preclean to remove food particles | Broom, dust pan, mop and bucket | ——— a.m. ——— a.m. ——— p.m. ——— p.m. and clean affected areas when spillages occur, without delay | Name |
| | 2 Run hot water directly into the clean mop bucket | | | Name |
| | 3 Add ——— mL of '———' detergent to hot water | '———' detergent | | Name |
| | 4 Mop the complete floor by frequently dipping the mop into detergent solution and applying adequate pressure | | | Name |
| | 5 Scrub difficult areas | Scrubbing brush | Prior to closing time After use | Name |
| | 6 Change water and detergent in bucket when dirty | | | |
| | 7 Empty bucket into gully trap adjacent to cleaner's room, rinse bucket and wash thoroughly | | | |
| | 8 Add ——— mL of '———' sanitiser into hot water drawn directly from hot water system and follow procedures 4 and 5 above | '———' sanitiser | | |
| | 9 Rinse mop in hot water, removing any debris. Wring out excess water from the mop. Hang mop to air dry. Store in cleaning cupboard. | | After use | Name |

Decontamination — cleaning and sanitising **61** ▶

**Table 11.2:** Personal cleaning and sanitising program

Name: _____

Week (dates): _____

| Monday | | Tuesday | | Wednesday | | etc. |
|---|---|---|---|---|---|---|
| Time | Job | Time | Job | Time | Job | |
| ____ a.m. | Check<br>• toilet paper<br>• hand towels<br>• liquid soap<br>in staff and public toilets | | | | | |
| ____ a.m. | Clean windows | | | | | |
| ____ a.m. | Dust and wipe shelving | | | | | |
| ____ a.m. | ………. etc. | | | | | |

When compiling a cleaning and sanitising program manually, you will need to list exactly:

- what is to be cleaned
- the procedure to be followed
- materials and products to be used
- how frequently performed and approximate time allotted
- person allotted to do the job (so that a job may not be neglected on the assumption someone else was doing it).

# Implementing a cleaning and sanitising program

## What is to be cleaned
Walk through your complete premises and list everything that requires cleaning at some time or another and overlook nothing. Start with the bare structure, then detail each individual room, its surfaces, fixtures, equipment and so on.

For the moment ignore your present methods and critically examine your needs with a fresh view. Ask your local environmental health officer for assistance and advice. They can explain the critical control points by which they normally check the effectiveness of your program to prevent food poisoning.

## How the job is to be done
Alongside each surface or article detail the exact procedure to be carried out; indicate any hazards that may be encountered.

## Equipment and materials necessary
List beside each individual procedure:

- what equipment and accessories are required
- what materials are to be used and at what strength for each particular surface.

## When the job is to be done
Beside each job, list when and how frequently the job should be carried out.

Where possible adopt a 'clean as you go' policy to prevent cluttering up your kitchen with soiled equipment and utensils, and to prevent food from hardening on the surface and bacteria from growing rapidly on the surface.

If anything cannot be cleaned and sanitised immediately after use, it must be cleaned before re-use.

Preferably you should give some guide as to the length of time required to carry out the job without prejudicing the quality of the job.

## Who is to do the job

Assign each job to a particular person and ensure that that person understands and can competently carry out these duties.

Generally workers are assigned to clean and sanitise the area they work in, preferably on a 'clean as you go' policy. Larger cleaning jobs may be shared or done on a rotational basis to prevent boredom and to share out unpleasant jobs.

## Implementation

These completed lists form your master cleaning and sanitising program. From this you can devise individual employees' cleaning schedules for each day of the week.

This method of making individual employees responsible for particular jobs is of value when staff operate in shifts, as jobs may be purposely left from one shift to the next and eventually neglected or forgotten.

Explain the programme to the staff and ask for opinions or improvements. Ensure that staff understand why they are carrying out a particular procedure and demonstrate the procedure if necessary.

# Supervision and evaluation

A manager's commitment to a program will be reflected in the commitment made by each individual employee.

The program will make new employees aware of the standards they are to uphold.

To evaluate the program, bacterial swabbing can be carried out and the results shown to the employees and tabulated on a graph.

One simple effective system of assessment is the **Diversey Monitor (bacterial) Sanitation Test Strips**. These strips are used to swab the surfaces, then sealed in plastic incubation pouches and incubated for 24 to 36 hours, using body heat by placing the strips in a pocket. Expensive incubators are not required. The strips are then matched against a colour chart which has a value of 1 to 5, indicating whether the effectiveness of cleaning and/or sanitising is:

- satisfactory
- borderline
- poor
- unsatisfactory
- contaminated.

# Pest control

## Common kitchen pests

Flies, cockroaches and rodents have always plagued people not only by causing tremendous food loss through consumption or spoilage but also by carrying disease. These pests inhabit highly contaminated places such as sewers and garbage areas and their feet and body hairs pick up pathogenic bacteria. In feeding on contaminated food, the pests' mouthparts become soiled with bacteria and their excreta contains these bacteria and viruses.

Insects and rodents will always be attracted to food premises and once they have established themselves, they multiply extremely rapidly. The more offspring they produce, the greater the damage, spoilage and risk of contamination (see Table 12.1). The stored food insect pests (including weevils, beetles and moths) can cause massive damage to foodstuffs, if food handlers ignore or do not recognise the adult insect. Some birds also feed on stored food, and their lice and droppings can foul areas.

## Control of insects and rodents in food premises

To keep premises free of these pests, food handlers need to do the following.

- Prevent their access into the premises and food storage areas.
- Deny them harbourage (shelter) if they do get in, by sealing off all cavities or hiding places.
- Deny them water and the food that attracts them by maintaining a high standard of housekeeping hygiene.
- Organise suitable eradication measures and maintain control to prevent their return.

A licensed pest control operator is the best person to carry out the eradication and implement a control programme. However, in the first place the food handler needs to **recognise the presence of pests** and to arrange for treatment. Good building construction and maintaining older premises in good repair will minimise access and harbourage, but

**Table 12.1:** Multiplication table for insects and rodents

| Pest | Average number of offspring | Lifespan | Approx. length of life cycle |
|---|---|---|---|
| Rats | 6–8 per litter 200 per year | 1–3 years | 22 day gestation |
| Flies | 6 batches of 50–150 eggs | Variable | • 1 day as egg<br>• 5 days as larva<br>• 5 days as pupa adult |
| Cockroaches: | | | |
| • German | 20 000/year | 14–26 weeks | 6–31 weeks |
| • American | 800/year | 15–84 weeks | 23 weeks |
| • Australian | 200/year | 17–26 weeks | 30–59 weeks |
| Stored food insect pests: | | | |
| • Rice weevil | 200–300 eggs | 2–3 months | 4 weeks |
| • Rust-red flour beetle | 1000 eggs | 7 months | 4 weeks |
| • Saw-toothed grain beetle | 375 eggs | several months | 3–4 weeks |
| • Tobacco beetle | 100 eggs | 2–3 weeks | 5 weeks |
| • Tropical warehouse moth | 250 eggs | 2 weeks | 5 weeks |
| • Indian meal moth | 200–400 eggs | 3 weeks | 4 weeks |

generally only those food handlers exhibiting high hygiene standards can limit attraction and thus maintain control.

To understand each pest we need to be able to identify the various stages in its development (life cycle) and its habits, and to recognise the signs indicating its presence.

## Rats and mice

### Habits

Rats and mice are nocturnal animals but will adapt their habits to suit the environment. Rats have poor vision and rely heavily on their highly developed sense of touch, hearing, smell and taste to guide them. They have coarse hairs on their body, protruding from their fur, which are extremely sensitive to touch. Like a blind person and their walking stick, they feel their way generally maintaining contact with a vertical surface, for example a wall. With this physical contact the natural oil in their fur brushes along the vertical surface. As rats habitually use the same runways a grease 'smear' is formed.

Their two large incisor teeth grow approx 13 cm per year and so they continually gnaw to keep their teeth short. They chew through a tough Queensland macadamia nut as if it were a piece of cheese. They also chew electrical wiring in ceilings causing electrical fires.

Mice have similar habits to rats. They, however, are attracted to cheese where rats generally are not. Mice are nibblers and cause massive damage going from one product to another. Mice are most active between 7.30 p.m. and midnight.

### Signs of presence

*Droppings*

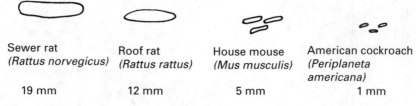

| Sewer rat | Roof rat | House mouse | American cockroach |
| (Rattus norvegicus) | (Rattus rattus) | (Mus musculis) | (Periplaneta americana) |
| 19 mm | 12 mm | 5 mm | 1 mm |

**Figure 12.1**

The size of the faeces indicates the pest species.

*Smears*
These indicate the runways the rats use.

*Characteristic odour*
Rats and mice have individual musty type odours.

*Gnaw marks*
The chisel-like marks of teeth are a giveaway. Rats and mice gnaw holes in walls etc. to gain access.

*Disappearance of food*
This, and damage to food, indicate the presence of rodents. Check for gnaw marks.

*Rat's urine*
This fluoresces under ultraviolet light. Mice produce urinating pillars which consist of dirt, grease and urine along the runs.

*Burrows/Nests*
The sewer rat prefers to burrow whereas the roof rat prefers to climb.

## Species
Sewer rat      *Rattus norvegicus* (25 cm body length excluding tail)
Roof rat       *Rattus rattus* (20 cm body length excluding tail)
House mouse    *Mus musculis* (9 cm body length excluding tail)

## Control

*Deny access*
In badly infested areas, doors may be metal-capped to prevent gnawing. Seal around pipes or ducts where they pass through the walls etc. Generally good building construction restricts access; however, to prevent access to the cavity area ensure that concrete masonry walls are capped. Otherwise the cavity may become 'high rise apartments' for the rat colony. Rat baffle walls should follow the perimeter of the building and extend 600 mm below the natural ground level. Check all sewerage/stormwater grates as rats will have direct access into a building from the sewers.

*Deny harbourage*
Eliminate their hiding places by keeping food storage items at least 30 cm above the floor, and away from the walls. This also makes it easy to clean up spilt items, limiting the amount of loose food around to attract rodents. The forgotten areas in store rooms, above cold rooms, sheds and rear yards accumulate unwanted materials that give excellent shelter to rats and mice.

*Deny them food and water*
Don't leave your cleaning until the morning as rodents feed overnight. All waste food should be stored correctly in tightly lidded metal refuse bins to prevent access to the food.

Good housekeeping reduces most problems. Rats may obtain their water from leaking taps or water ponding in roof guttering or from the condensate from airconditioning units. Eliminating this water and asking the pest control operator to provide poisoned water will give effective results.

## Eradication

### Trapping
Use back-breaking and cage traps.

### Poisoning
This should be left to licensed pest control operators. However, anti-coagulants such as Ratsak and Racumin should be kept on hand as an insurance. Rats need at least five consecutive days feeding on anti-coagulants before they die, so as long as they are eating it, keep feeding it to them.

# Flies

### Habits
The fly does not have teeth to chew with but has sponge-like mouthparts and has to have its food in a liquid form. To consume solids, it firstly regurgitates a liquid from its stomach onto the food. This liquid dissolves the solid food and the fly then mops up the liquid with its sponge-like mouthparts.

Flies may feed off garbage, animal manure etc. and regurgitate the liquid containing large numbers of pathogenic bacteria onto our food or preparation surfaces. The bacteria may also be found on the flies' feet, body hairs and in their faeces.

Flies usually dislike rapid air movement and find sheltered resting sites (usually horizontal surfaces).

### Signs of presence
The various stages in the life cycle: eggs, larvae (maggots), pupae, adult flies, fly specks (droppings) on the ceiling etc.

### Species
The housefly *Musca domestica*
Blowflies including *Lucilia cuprina* (sheep blowfly)

### Life cycle
egg — larva — pupa — adult

From egg to adult under Australian summer conditions may be as short as nine days for the housefly.

The small elongated eggs are creamy white and laid in batches of up to 50 in most animal manures, decaying food and mulch. The egg hatches within a day to the larval stage. The white larva (maggot) is legless but moves quite readily and actively partakes of food in warm moist conditions. The larva then pupates into a motionless brown pupa case, having sought a drier cooler position. Some larvae crawl out of garbage bins and pupate in the soil or area adjacent to the bins. The adult hatches from the pupa case and after drying its wings for an hour or so is capable of flying.

Blowfly species go through the same cycle. However, in summer the female adult blowfly may skip the egg stage and actually lay the lavae directly onto food.

*Lucilia cuprina*, initially responsible for the primary blowfly strike in sheep, has now turned its attention to striking garbage and is a major problem in Queensland.

## Control

### Deny access
Use fly screening or air curtains to prevent access. Airconditioning a building also prevents access as doors and windows have to be kept closed.

### Deny food
Promptly remove waste food to an adequate number of plastic-lined garbage bins with tight-fitting lids. Cooled or refrigerated storage rooms reduce odours and lessen the attraction for flies, and daily or multiple garbage removal services per week also assist in controlling a fly problem.

### Eradication
- Spray residual sprays on resting sites (usually horizontal surfaces).
- Use knockdown (or space) sprays to kill the flies present at that particular time. One squirt is sufficient — don't drown the fly as it usually takes up to 5 to 10 minutes to die.
- Use electronic insect killers. However, position them so that the preparation area is not bespattered with the incinerated bodies of the electrocuted insects.
- Use fly baits around garbage areas.

If you are having problems with fly infestations or the presence of fly larvae (maggots) ask your local environmental health officer for advice.

# Cockroaches

## Habits
Cockroaches are gregarious nocturnal (night) insects that shun light, consequently they inhabit kitchens when we turn off the lights and leave.

During the day they rest in warm moist dark areas such as sewers, refuse areas, under refrigerators, stoves, hot-water systems, sinks and cupboards, in cracks and crevices and also in packaging, for example drink and food cartons and sacks of potatoes.

Because they inhabit sewers, their legs, body hairs, mouthparts, saliva and faeces, as well as the food they regurgitate, carry pathogenic organisms (for example *Salmonella*) and contaminate food and surfaces.

## Signs of presence
Egg capsules, nymphs, adults, droppings, characteristic odour and taints, signs of damage through eating.

## Species
The German cockroach, *Blatella germanica* (small, light honey/brown cockroach); 20 mm body length

The Australian cockroach, *Periplaneta australasiae* (large, reddish-brown cockroach); 32 mm body length

The American cockroach, *Periplaneta americana* (very large, reddish-brown cockroach); 50 mm body length

The Oriental cockroach, *Blatta orientalis* (large, dark brown/black cockroach); 35 mm body length

There are over 3500 species.

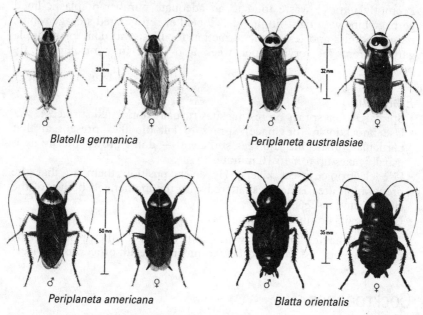

**Figure 12.2** Cockroach species

## Life cycle
egg — nymph — adult

The egg capsule (ootheca) contains a number of eggs in several rows. The female cements the egg capsule in a suitable position. The eggs hatch into nymphs (immature, wingless cockroaches) that undergo several moults and develop into mature adult cockroaches with fully developed wings.

## Control

*Deny access*
Provide close-fitting doors to the premises.

*Deny harbourage*
Maintain good building practices and verminproof the complete premises by sealing off all cracks and crevices to eliminate harbourages.

*Deny food and water*
- Clear food waste away into plastic-lined garbage bins with close-fitting lids.
- Clean up kitchens before going home of an evening (cockroaches are nocturnal feeders).
- Ensure that all food is kept in vermin-proof containers.
- Ensure that all crockery etc. is kept in vermin-proof cupboards.
- Make sure there are no water leaks to provide moist conditions in a warm kitchen, for example under sinks and hot-water systems.

## Eradication

Undertake residual spraying of the complete premises concentrating on the areas inhabited by cockroaches (see habits).

A licensed pest control officer is equipped with chemicals, equipment and safety gear to carry out this control. Space or knockdown sprays, baits and sterilisation chemicals can also be used by food handlers when cockroaches are noticed.

# Stored food insect pests

## Habits
These insects are very small. The adults lay their eggs on the packets or in spilt food, and the hatching larvae chew through the packets into the food. Larvae leave an easily noticed web that binds the food particles together in clumps.

Moths tend to shun light and to rest in areas under shelves, whereas beetles tend to move around on top of the shelves. These insects require warm humid condition for ideal growth. Cooler dry conditions slow down the life cycle. Eggs may be laid in foodstuffs at any point along the food chain from manufacture to storage in the final retail outlet. Under ideal growth conditions they continue with their cycle and adults are likely to hatch out in the food premises.

## Signs of presence
- Holes in packets.
- Presence of larvae and associated web, binding the food particles together.
- The presence of pupae on the folds of the seams on plastic and cellophane packets. The larvae also pupate in the folds of the lid and corners of cardboard packets.

- The presence of adults in the food packages, or under shelves, or flying freely about in a slow, irregular flight.

## Species
The rice weevil *Silophilus oryzae*
The rust-red flour beetle *Tribolium castaneum*
The saw-toothed grain beetle *Oryzalphisus surinamensis*
The tobacco beetle *Lasioderma serricorne*
The tropical warehouse moth *Ephestia cautella*
The Indian meal moth *Plodia interpunctella*

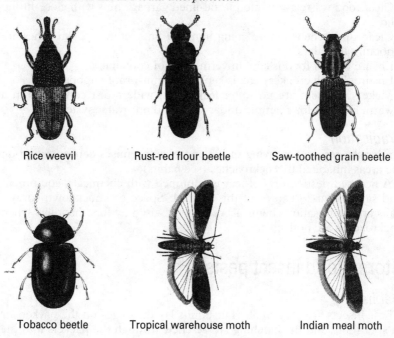

**Figure 12.3** Weevils, beetles and moths

## Life cycle
egg – larva — pupa — adult
The length of the life cycle varies with the different species of pests (see Table 12.1).

Temperatures above 21°C and relative humidity above 10% favour growth.

## Control

### Deny access
Inspect foodstuffs coming into the premises for these insect pests as their short life cycles and high reproductive rates may cause untold

damage to existing stock. Refuse to accept infested stock and advise the distributor and/or manufacturer of the reasons for refusal. This should prevent further contaminated deliveries, as reputable firms would locate the problem and correct it.

### *Deny harbourage*
Clean up any spilt foods on shelving or in cracks and crevices where shelving adjoins the wall. In some cases ill-fitting 'price marking' capping at the front of the shop shelves traps spilt food.

Spacing the shelves 25 mm away from the wall allows food to drop and be cleaned up during a regular cleaning program.

### *Deny food*
If products are to be stored for considerable periods, place them in insect-proof containers with tight-fitting lids. In slow-selling products the insects have time to go through their life cycle on the premises and reinfest existing stocks. Rotating stock and keeping only minimal quantities are important to limit this storage time.

Unless poultry or bird feed are fast sellers, retail stores selling these or similar products should limit the quantity they store to get a quicker turnover. These items invariably arrive already contaminated and reinfest the premises.

## Eradication
- If insect infestation is noted, remove the infected food to the refuse tip *immediately*. Check for adult insect infestation and spray with a *knockdown or space spray* to prevent further infestation.
- Clean down the shelving, and spray it on both sides with a *residual spray*. (Remember that moths usually are located under the shelving.)
- Check the food items as they are put back on the shelf. Shake any products packed in clear cellophane or plastic to bring any adult beetles and moths to the surface where their movements can be noticed.
- Keep a small space (approx. 25 mm) between each different *type* of product stored. Beetles usually will migrate to this space and can be easily noticed. When the items are stocked close together the insects pass from one type of product to the next unnoticed.
- If the infested product has to be kept in connection with a claim, store in an airtight container and spray with a surface spray. Insect-infested food is unsuitable for use.

# Kitchen design and construction

## Design

Growth in the tourism and hospitality industry has placed larger demands on the food service manager and their staff to deliver a fast customer service with safe, wholesome, inviting food. If a food establishment is correctly designed it can streamline work practices, reduce cleaning and sanitising costs and prevent cross-contamination of foods.

The environmental health officer sees the benefits and pitfalls in the various designs, equipment and materials used in certain situations and can offer practical solutions to problems encountered in this fast changing food industry.

## Layout

The first consideration in kitchen design is a layout that gives a **logical continuous work flow**. To do this, the design should separate the preparation and serving areas from contaminable areas, for example those receiving raw produce, those receiving soiled articles from a dining room, refuse storage areas, washing up areas and office and staff facilities.

- Raw produce, for example potatoes, are generally heavily contaminated with soil. If the storage area is close to the delivery door, delivery personnel do not need to tramp through the preparation area, possibly soiling and contaminating it.
- Washing up areas receive soiled crockery and cutlery plus waste food. They should be well separated from preparation and serving areas so that staff are not tempted to contaminate benchtops by temporarily stacking soiled items on them until ready for washing.
- Cooked food and raw food areas should be separated from one another to limit cross-contamination either in preparation or storage.
- Clothing and other non-food items, for example cardigans etc. should be kept out of preparation areas. Delivery dockets and correspondence stacked in preparation areas can also cause contamination. Staff and office areas should be physically separated from the other areas.
- Store refuse away from the kitchen facilities.
- Contaminable items such as brooms and mops should be stored out of the preparation areas and not left lying in the corner.

Kitchen design and construction **75**

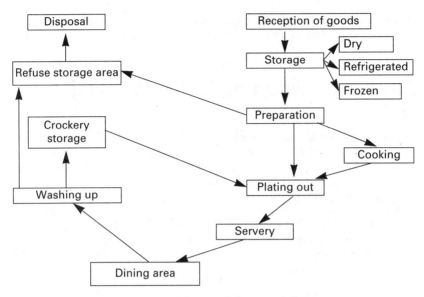

**Figure 13.1** Kitchen layout providing continuous work flow

## Cleanability

The second consideration is the **cleanability** of each area, its surfaces and equipment. If items are easy to clean, they will retain an 'as new' appearance and staff will display enthusiasm and pride in maintaining a high standard of cleanliness. On the other hand, if a surface is difficult to clean, the drudgery involved will produce frustration and the feeling that the attempt is a waste of time. Food-poisoning bacteria love these neglected areas as it gives them *time* to grow to the infective dose in between each cleaning and the ability to move directly from surface to food.

Placing equipment on castors enables easy cleaning of the areas around, under and behind the equipment.

## *Construction*

If you plan to establish food premises in an existing building, check that your proposal meets Town Planning and Building Department requirements. If the basic requirements are fulfilled, you will need to submit detailed plans to the Local Authority Health Department stating:

- a menu or list of foods to be prepared or sold. This indicates the type and scale of operation proposed and what specialised surfaces you propose to use. A HACCP food safety plan may be required
- the floor plan and sections listing a specification of materials to be used in the construction
- the type and position of all equipment to be used.

## Structural requirements for food premises

### Size
The building must be adequate in size so that all processes can be carried out *within* the building. Food cannot be prepared in private household kitchens, but must come from authorised food establishments and be delivered in approved delivery vehicles.

Check with your Local Authority Health Department for minimum space requirements and the relationship between kitchen and dining space for a specified number of patrons. Building, Liquor Licensing and Fire Prevention Departments also make specific space requirements per patron.

A cramped kitchen restricts work flow, reduces productivity and generally doesn't allow easy access for cleaning.

### Separation of areas
Kitchens and preparation/cooking areas need to be physically separated from the public sector to prevent the possibility of droplet spread of bacteria and viruses through coughing and sneezing.

### Floors
Floors should have a smooth (yet non-slip), continuous, impervious, hard-wearing surface so that they are capable of being easily cleaned and sanitised. The floor covering must suit the purpose of the process carried out in the area. In smaller establishments where little dry spillage occurs, a moppable surface such as welded sheet vinyl would suffice. In larger establishments where liquid or heavier spillages may occur, washdown facilities are required and the floor must be graded (slope) to a floor waste drainage outlet.

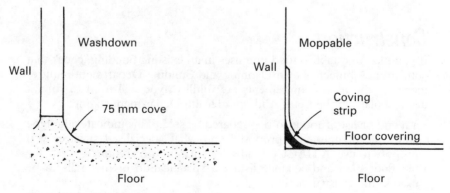

**Figure 13.2**  Washing floors

Around the perimeter of the room at the junction between the floor and the walls or fixtures, the floor is coved up the wall with a curve of radius of 75 mm.

## Walls

Walls should have a smooth continuous surface which is capable of easy cleaning. In preparation areas where little splashing occurs, a light-coloured gloss paint would suffice. Where heavy splashing occurs or where washdown facilities are required, the walls should be lined with a more durable material such as stainless steel or ceramic tiles with epoxy grout.

The junction of the walls should have a 25 mm radius cove to eliminate corners and facilitate easy cleaning.

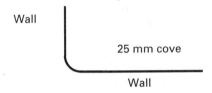

**Figure 13.3** Washing walls

## Ceilings

As for walls, ceilings should be smooth and continuous with a light-coloured gloss paint. Suspended ceilings are difficult to clean and if not fitted correctly can harbour vermin and allow debris to drop down from the ceiling. The minimum ceiling height is 2.4 m.

## Light

Adequate light (natural or artificial) is required for efficient working conditions and to allow inspection for cleanliness. The Australian Standard (AS 1680–1976) can be used as a guideline. Flush-mounted lights enable easy cleaning.

## Ventilation

Adequate natural or artificial ventilation is required to ensure efficient working conditions. Poor ventilation can make a kitchen with cooking appliances unbearable to work in and produce temperatures ideal for bacterial growth.

Airconditioning or a mechanical air supply system can provide ventilation where natural ventilation is inadequate.

## Preparation benches

These surfaces should be non-toxic, hard-wearing, free from cracks and crevices and easy to clean and sanitise.

Each surface should be suitable to its particular use, for example on those subjected to heavy use, slight impacts or heat, stainless steel is more suitable and durable than laminated plastics.

Splashbacks to the bench should be a minimum height of 300 mm and in the case of stainless steel, incorporated into the bench to eliminate a join. Chipped, scratched, cracked or damaged surfaces can retain food particles and juices as well as food-poisoning bacteria. Timber-constructed benches and cutting boards should be prohibited as they are incapable of being thoroughly cleaned and sanitised. Removable dense-polythene benchtops and cutting boards can be easily cleaned and sanitised.

## Handwashing facilities
There should be handbasins in the food preparation area and they should be used solely by food handlers for washing their hands. There should also be handbasins in the toilet area for when staff have visited the toilet. The use of handbasins in these two areas should be kept quite separate.

There should be an adequate number of handbasins for the number of employees, provided with hot and cold water, liquid soap, nail brush and disposable hand towels or air dryers. Automatic sensory devices eliminate the need for taps (hand, elbow or foot operated), reducing cross-contamination of hands, and enabling easy cleaning of adjacent walls and floor.

## Dishwashing facilities
Automatic dishwashers will give better results than dishwashing by hand in a double bowl sink. With sinks the first bowl is designed for cleaning, the second bowl for sanitising. An adequate supply of hot and cold water is required to carry out these two processes, several times a day. Hot-water systems are located external to the kitchen.

The sink bowls must be of an adequate size to take the largest equipment. If an item is too large to fully immerse for sanitisation by heat, adequate facilities should be available for chemical sanitisation and rinsing.

## Cooking appliances
These are either free-standing or bench-mounted. Most bench-mounted items are easily removed for cleaning of the bench and splashback.

If items are too large to shift (as most free-standing appliances are) they need to be either:
- placed on castors and have flexible power or gas leads to allow them to be pulled out so they can be cleaned under and around, or
- flashed and sealed to the splashback so that they leave no harbourage for vermin.

## Exhaust ventilation
Any smoke, fumes, vapours, grease, heat or odours given off by the cooking appliances are collected, filtered then discharged by means of an exhaust hood and ventilation system mounted above the cooking appliances.

Such exhaust ventilation must comply with Australian Standard AS 1668, Part 2.

## Refrigerators/Freezers

Refrigerators, cold rooms and freezers shoud be able to maintain temperatures below 5°C for refrigerators and below −18°C for freezers, even if the doors are constantly opened and shut, as they are in our Australian summers.

Motors or compressors should be located externally where possible, unless they are an integral part of the appliance. If mounted internally, they must be protected against spillages and included in the maintenance programme. Internal mounting may reduce the efficiency of the appliance and also increase the ambient room temperature.

In the cold room there should be adequate shelving of a corrosive-resistant, non-absorbent material to keep food above the floor to enable easy cleaning. Other provisions include condensate drainage and temperature gauges.

Free-standing refrigerators should be mounted on castors to enable easy cleaning.

## Food displays

Where food is displayed to the public it should be adequately protected against droplet spread by wrapping the product. The display cabinet should be protected by a sneeze guard. Potentially hazardous food should be either refrigerated below 5°C or kept hot above 60°C to prevent bacterial growth.

## Refuse storage and disposal

Refuse and waste food in garbage bins generally remain at room temperature resulting in high bacterial counts. The bins should be emptied regularly and not allowed to overflow. Handwashing after handling garbage is also important. In larger establishments, the internal storage bins should be lined with a plastic bag, which when full should be tied and sealed and put into the external storage bins.

The external garbage bins should be located to allow easy access for collection and away from any airconditioning intake, mechanical air intake or window. This area should have washdown facilities with the floor sloping to drain into the sewerage system.

The bins should be of an approved vermin-proof and fly-proof construction with adequate capacity for the number of services per week.

## Storage for cleaning equipment

Cleaning equipment is by far the most contaminating equipment used in the kitchen and should not be left lying around the kitchen area. A room or cupboard should be provided for detergents, sanitisers and cleaning equipment. Never store these items above a preparation bench as spillages may result in chemical food poisoning. The mops should be sanitised, dried then stored away. A slop sink or disposal point should be located so that the floor washings can be emptied into it. **Never empty floor washings into a sink or handbasin.**

# Food poisoning in the home

Food poisoning is less prevalent in the home because generally:
- food is cooked to be eaten almost immediately
- only sufficient servings for each particular meal are cooked, not allowing bacteria sufficient time to reach the infective dose or produce spores or toxins.

## Food hygiene
Problems can and do arise through the following.

### Bulk cooking
Home catering for a birthday party or other function with **inadequate facilities** for preparation, cooking and/or refrigeration can lead to contamination and bacterial growth.

### Use of leftover foods
It can be dangerous to use leftover food that has been sitting for some days in a fridge whose temperature has been fluctuating.

### Leaving food at room temperature
It is also an undesirable practice to leave food in the danger zone for a considerable time in order to:
- *thaw*, for example leaving frozen food in the kitchen sink or on the bench (see chapter 10, Frozen food)
- *cool*, for example leftovers from the evening meal put aside before refrigerating or freezing. The roast is often left out for hours after carving or until the table is cleared for washing up.

### Transporting food
- **Keeping frozen food in the boot or interior of your car while you do your shopping is dangerous.** Even if the car is airconditioned the food will still be in the high danger zone. It is best to buy frozen foods just before you go through the checkout, then transfer them to a chilled Esky in your car. This not only prevents food poisoning but retains the quality and flavour of the food.

# Food poisoning in the home

- If you have to transport hot or cold food to a family gathering, preheat or precool an Esky using hot water or ice respectively, before putting the sealed hot or cold food containers in it.

> Thaw, cook and serve food immediately to minimise the time in the danger zone and restrict bacterial growth.

## Faecal contamination

This can occur through the presence of the family pet in the kitchen or young toddlers in nappies.

It poses a problem when the person at home:

- handles a pet or changes a nappy without thoroughly washing their hands afterwards
- washes their hands in the kitchen sink
- wipes their hands on a tea towel.

## Kitchen equipment

The domestic kitchen is built to a less stringent standard of design than that of a commercial kitchen and different factors have to be taken into account.

### The domestic refrigerator

This is designed to cope with less frequent door openings than a commercial refrigerator. Children stand and make long decisions on what

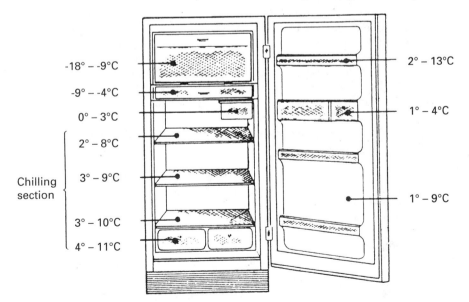

**Figure 14.1** A domestic refrigerator

they want, with the refrigerator door wide open, allowing cold air to escape and hot air to enter. Some children also stand in front of the open refrigerator for relief from the summer heat. With the door open for a one minute period in summer, the air temperature in your refrigerator may rise from below 5°C to approximately 14°C. Check the temperature range for the various positions for your refrigerator. Generally it will be similar to that in Figure 14.1. Allow the air to circulate by not overpacking.

Stocktake now and then and remove forgotten leftovers. Refrigerate leftovers within 15 to 20 minutes as modern refrigerators can cope with receiving food that is still hot. If you have odours in a refrigerator, remove the offending item and wash and sanitise the interior. Odoureaters (activated charcoal) can be placed in the refrigerator to absorb odour or the surfaces can be wiped over with a damp disposable cloth and vanilla essence.

## Freezers

The upright freezer looks nicer than the chest-type freezer but loses cold air quicker when the door is opened. Arrange food to allow air to circulate around it. Freeze food in portions rather than in bulk to prevent the unhealthy practice of thawing and refreezing the unwanted portion. If a food item has been partly thawed in a refrigerator or microwave it *may be safely refrozen* as long it has been kept out of the danger zone (below 5°C).

## The microwave oven

Microwaves vibrate the water molecules in food producing heat. Microwaves are excellent to thaw food quickly to 5°C.

Food needs to be subjected to adequate temperature for a certain contact time to destroy bacteria. Microwave ovens cook food extremely quickly and evenly; the food should reach the desired temperature and be left to stand for a period to ensure heat is conducted to the drier areas of the food. Whereas food in a conventional oven has to cook until heat reaches its centre, microwaved food may be cooler on the surface and hotter in the centre, as heat is radiated into the surrounding air space of the microwave oven from the food surface.

Covering food not only retains the flavour but also the moisture (and heat) for the microwaves to work efficiently. Ensure that you use a microwave-safe wrapper.

Keep the interior of the microwave oven clean as swabbing has indicated bacteria may survive on the cold microwave oven surfaces. This is possibly because the bacteria are subjected to inadequate heating, whereas the heat on the conventional oven walls kills most bacteria.

## The kitchen sink

This should not be used for washing hands (causing contamination) or for thawing food (causing cross-contamination) as the sink is generally highly contaminated from the preparation of food, for example washing vegetables and salads etc.

## Food covers
Cover food with a disposable wrapper, *not* an absorbent cloth or tea towel. Separate raw from cooked foods and store cooked above raw to prevent cross-contamination by contact or 'drip' respectively.

## Cutting boards
Throw away your timber cutting board and replace it with a white dense-polythene cutting board that you can sanitise. It can be put in your dishwasher or have bleach, such as Sno White, poured over to sanitise the surface. Bleach not only sanitises it but keeps the board white and free from putrid smells.

## Tea towels
Use tea towels sparingly and correctly. If you have a dishwasher there is no need to dry up. If you only have a single bowl sink either *soak* dishes in hot water while washing up or pour very hot water over them in the drainer and allow to *air dry* using the heat they have absorbed.

Cross-contamination occurs when the tea towel is used for drying hands, or as an all-purpose cloth for drying tables, benches and surfaces, or for covering food then for drying up.

## Benchtops
Soiled or stained benches can be easily cleaned using detergent, baking powder or creme cleaners and easily sanitised with methylated spirits on a clean cloth.

## Cleaning cloths/Scrubbing brushes/Scourers
Cleaning cloths can cross–contaminate surfaces. Cleaning aids should be thoroughly cleansed of food debris, sanitised by standing in hot water for a few minutes then opened out to air dry.

# Appendix 1
# Ciguatera

Ciguatera is a natural food poisoning prevelant in the tropical and subtropical areas of Australia caused by the consumption of large reef fish such as Spanish mackerel, snapper, coral trout, groper, parrot fish, reef cod, barracuda and surgeon-fish, which contain ciguatoxin. Although further research is still being carried out, there is strong evidence that ciguatoxin is produced by tiny dinoflagellates, *Gambeirdiscus toxicus*, that live in algae found on coral in the Great Barrier Reef. Small herbivorous fish eat the algae, along with the dinoflagellates and their toxin, and the toxin accumulates in these small fish. Larger predatory fish eat the smaller fish and the ciguatoxin accumulates in the skin, muscle and viscera of the large fish.

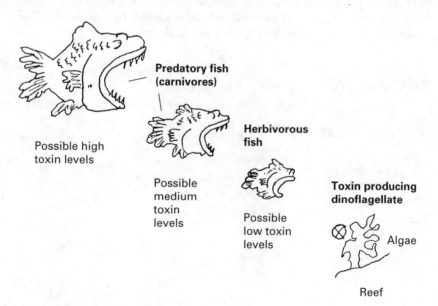

**Figure A.1**  A ciguatera chain

The ciguatoxin cannot be detected by smell, taste or appearance of the fish. The toxin can withstand freezing, drying or salting and is heat-stable.

Ciguatera cases are sensitive to any further very minute levels of ciguatoxin, so they must not eat reef and offshore fish. Alcohol consumption may aggravate symptoms or cause neurological symptoms.

## Prevention of the disease

1. Avoid fish such as red bass, chinaman fish, and paddle tail which are 'high risk' fish species, or other suspect reef fish weighing over 10 kg.
2. Choose reef fish species with total weight of less than 3 kg where possible.
3. To reduce susceptibility prepare only small portions of fish for each serve and don't use the same fish for repeated meals.
4. Never eat the roe, intestines, liver and other viscera of reef fish.
5. No practical laboratory test for ciguatoxin is available. If a large reef fish of up to 10 kg is required, test for the presence of the toxin by feeding sample pieces of the fish to a cat. Look for symptoms of unco-ordination, paralysis in the hindquarters, or vomiting and diarrhoea within six hours.

# Appendix 2
# Incubation period, duration of illness and major symptoms of food poisoning

| Food poisoning | Incubation period (in hours unless otherwise stated) | Duration of illness (days) | Major symptoms (characteristic symptoms in bold) |
|---|---|---|---|
| *Salmonella* | 6–72 (av. 18) | 2–7 | **Fever, diarrhoea,** abdominal pain, vomiting **Can be fatal** |
| *Staphylococcus aureus* | 1–6 (av. 4) | 1 | **Vomiting, hypotension,** diarrhoea, abdominal pain |
| *Clostridium perfringens* | 8–22 (av. 14) | 2 | **Diarrhoea, abdominal pain** |
| *Vibrio parahaemolyticus* | 2–48 (av. 12) | 2–5 | **Profuse diarrhoea,** vomiting, abdominal pain |
| *Bacillus cereus* | | | |
| • Vomiting type | 1–5 | 1 | **Vomiting,** nausea, diarrhoea |
| • Diarrhoea type | 8–16 | 1 | **Abdominal pain, diarrhoea,** nausea |
| *E. coli* | 5–48 | 1–5 | **Abdominal pain, diarrhoea,** vomiting, fever, chills, headache, muscular pain |
| *Shigella* spp. | 24–72 | 7 | **Abdominal pain, diarrhoea (with blood and mucus), fever** |
| *Campylobacter fetus* | 2–5 days | 2–3 | **Diarrhoea with blood and mucus, abdominal pain,** nausea and fever |
| *Yersinia enterocolytica* | 24–36 | 7–14 | **Fever, appendicitis-type pain, dysentery** |

| Food poisoning | Incubation period (in hours unless otherwise stated) | Duration of illness (days) | Major symptoms (characteristic symptoms in bold) |
| --- | --- | --- | --- |
| *Listeria monocytogenes* | 1–21 days usually 2–3 days | can be prolonged | **Flu-like symptoms including fever.** Nausea, vomiting, headache, causes miscarriages and still births, and meningitis |
| *Clostridium botulinum (E)* | 12–36 | 3–10 | **Fatigue, headache, dizziness, visual disturbances, inability to swallow, paralysis Generally fatal** |
| Viral gastroenteritis | 24–48 | 1–2 | **Nausea, fever,** diarrhoea and vomiting |
| Ciguatera | 1–36 (av. 5–12) | Generally 3–4 (Some symptoms may persist for years) | Nausea, vomiting, diarrhoea, abdominal pain followed by **numbness and tingling of the extremities and temperature reversal sensations** |

# Appendix 3
# Emerging pathogens causing foodborne illnesses

| | |
|---|---|
| **Traveller's diarrhoea** | (Infection or intoxication depending on strain) |
| Causative agent | *Escherichia coli* (*E. coli*) (Several strains, some pathogenic, others harmless) |
| Habitat | Intestines of humans and animals |
| Foods involved | Most foods and water |
| Mode of transmission | Cross-contamination from infected food handler |
| Control measures | 1  Good personal and food hygiene practices<br>2  Adequate cooking and storage temperatures (hot and cold)<br>3  Satisfactory cleaning and sanitising of surfaces and equipment. |
| | |
| **Bacillary dysentery** | (Infection-intoxication) |
| Causative agent | *Shigella* spp. |
| Habitat | Intestines of humans, insects and rodents |
| Foods involved | Salads (including potato, macaroni); milk and dairy products; water |
| Mode of transmission | Infected food handlers and carriers |
| Control measures | 1  Good personal and food hygiene practices<br>2  Adequate cooking and storage temperatures (hot and cold)<br>3  Thorough pest control. |
| | |
| **Campylobacteriosis** | (Possible infection and intoxication) |
| Causative agent | *Campylobacter fetus* subsp. *jejuni* |
| Habitat | Intestines of animals and birds (human carriers up to 1 year) |
| Food involved | Raw meat and poultry; raw milk and cream |
| Mode of transmission | Cross-contamination |

| | |
|---|---|
| Control measures | 1 Good personal and food hygiene practices<br>2 Satisfactory cleaning and sanitising of surfaces and equipment<br>3 Thorough cooking as the bacteria are heat-sensitive. |
| **Yersiniosis** | (Infection or intoxication depending on strain) |
| Causative agent | *Yersinia enterocolytica* (Several strains, some pathogenic, others harmless)<br>Infective dose approx. 10 micro-organisms/g of food |
| Habitat | Decaying organic matter, dust, water, humans, animals |
| Foods involved | Milk and dairy products; poultry products; pork products |
| Mode of transmission | Cross-contamination by food handlers or surfaces. Heat readily kills.<br>Bacteria however *can grow at 0°C* (refrigeration temperatures) |
| Control measures | 1 Good personal and food hygiene practices<br>2 Thorough cooking<br>3 Satisfactory cleaning and sanitising of surfaces. |
| **Listeriosis** | (Infection) |
| Causative agent | *Listeria monocytogenes* |
| Habitat | Soil water and plants; intestines of humans, birds and animals. Grows slowly at 0° to 3°C |
| Foods involved | Vegetables grown in soil; dairy products, including soft cheeses; raw meat and poultry; prepared, chilled, ready to eat foods |
| Mode of transmission | Cross-contamination of moist environments, cleaning equipment |
| Control measures | 1 Thorough cooking<br>2 Prevention of cross-contamination. |
| **Botulism** | Intoxication |
| Causative agent | *Clostridium botulinum (E)* |
| Habitat | Spores found in soil and animal intestines |
| Foods involved | Low acid canned products including fish and vegetables that have been inadequately heat-processed. Spores not killed.<br>Generally home-processed foods; honey. |

| | |
|---|---|
| Mode of transmission | The bacteria produce a toxin which is fatal. Bacteria grow in canned foods stored at room temperature for long periods<br>Honey may contain spores |
| Control measures | 1 Adequate heating (toxin destroyed by heat)<br>2 Check for blown, deformed or defective canned foods; **do not use or even taste**<br>3 Do not feed honey to babies under 12 months of age. |

# Appendix 4
# Critical control point decision tree

(Apply at each step of food preparation with an identified hazard)

Q 1. Do preventive measure(s) exist for the identified hazard?

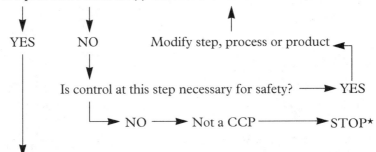

Q 2. Does this step eliminate or reduce the likely occurrence of a hazard to an acceptable level?

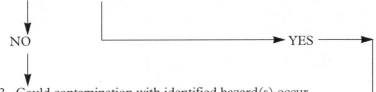

Q 3. Could contamination with identified hazard(s) occur in excess of acceptable level(s) or could these increase to unacceptable level(s)?

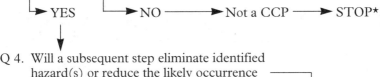

Q 4. Will a subsequent step eliminate identified hazard(s) or reduce the likely occurrence to an acceptable level?

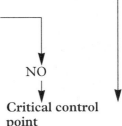

*Proceed to next step in the described process

# Index

aerobes 16
alcohol 9, 43, 58
anaerobes 16, 28
antibiotics 2, 18

bacillary dysentry (see *Shigella*) 88, 86
Bacilli 4
*Bacillus cereus* 4, 7, 12, 21, 23, 29–31, 37, 41, 42, 86
bacteria
  as a cause of food poisoning 3
  destruction of 6, 14, 57, 58
  growth of 4–5, 49
  habitat 23, 26, 28, 30, 31, 88–90
  movement of 4
  reproduction of
  shape 4
  size 3
  structure 4
  usefulness 3
bacterial food poisoning 11, 21, 22, 23
bain marie 23, 55, 79
birds 23, 28, 64
body habits 26, 33, 35, 54
blown cans 50
bulk cooking 28, 38, 80

*Campylobacter fetus* 86, 88–9
canned foods 18, 50, 89–90
carrier 23, 25, 26, 33, 34
causes of food poisoning 2, 22
ceilings, structure 77
cell structure 4–5
chemical food poisoning 20
ciguatera 84–85, 87
cleanability 75, 77, 83

cleaning
  of kitchen 57–63, 79
  program 59–63
  and sanitising 57–63
  of surfaces 2, 83
Cleaning Master 48, 59
*Clostridium botulinum* 40, 41, 87, 89–90
*Clostridium perfringens* 7, 21, 23, 27–9, 36, 37, 38, 39, 42, 86
cocci 4
cockroaches 65, 69–71
coldroom 13, 31, 51, 52, 79
contamination 5, 11, 14, 15, 16, 22, 24, 25, 26, 27, 28, 29, 32, 35, 36, 37, 38, 39–43, 49, 50, 51, 52, 54, 55, 57, 58, 59, 64, 69, 74, 82, 83, 88, 89
cooking of food 9, 13, 14, 25, 29, 30, 32, 38, 54, 56, 74, 78, 80, 82, 88, 89, 90
cooling of food 13, 14, 17, 29, 30, 38, 51, 79, 81–2
commandments for food poisoning 2
critical control points 44, 45, 47
cutting boards 25, 54, 83

danger zone for food poisoning 12, 13–14, 22, 30, 31, 38
decontamination 57–63
dehydrated food 17, 50
dehydration 17
detergents 58, 60
displaying of food 55
disposable gloves 35, 54, 55
drying
  of crockery 59
  of hands 36
  of surfaces 59

duration of illness   2
dysentery, bacillary (see *Shigella*) 86, 88

eggs
    cockroaches (ootheca)   70
    chicken   24, 40
    duck   24
    fly   65
    stored food insect pests   65, 68, 69
emerging pathogens   88
environmental conditions   13
environmental health officer   50, 62
*Escherichia coli*   41, 86, 88
evaluation   37–38, 63

factors that affect bacterial growth   11–16
faecal contamination   24, 28, 34, 36, 40, 66, 68, 70, 81, 88, 89
flies   68–69
floors   76
foodborne illness (food poisoning)   1, 8, 20–1, 22–32, 86–90
food displays   79
food handler   1, 22, 25, 26, 27, 33–6
food hygiene   49
food poisoning
    from bacteria   1, 21, 22–32, 86–90
    causes of   22
    from chemicals   20
    duration of illness   2, 86
    factors affecting   11–16
    from mycotoxins   10, 21
    from natural sources   20
    symptoms of   86–87
    from viruses   21
food premises   74–79
food safety (HACCP) plans   44–8
food safety legislation   44
food spoilage   3
freezers   79, 82
frozen foods   14, 17, 32, 52, 53
fungi   9–10

growth of bacteria   5–7, 11–16

HACCP   44–8
HACCP Master   47–8
hands
    drying   36
    as a source of infection   25, 26, 27, 28, 32, 34, 35, 36, 39, 41, 54, 55, 56, 78, 81, 88–9
    washing   35, 36
Hazard Analysis, Critical Control Points   44–8
healthy carrier   23, 25, 26, 33, 34
heat treatment of food   13–14, 17, 25, 27, 28, 29, 30, 31, 32, 38, 54, 56, 80, 82, 88–90
hepatitis A   21
humidity   16

incubation period   2
infections   8, 23
infective dose   22, 23
insects   64, 68–73
intoxication   8, 23
irradiation   18

jewellery   35

kitchen design   74–9

leftovers   56, 80
lighting of kitchen   77
*Listeria monocytogenes*   87, 89

metho (methylated spirits)   58
mice   65, 66–68
micro-organisms   3–10
microwave oven   52, 82
moisture content (see water activity)   11
mould   9, 12, 21
movement   5
mycotoxin   10

nails   35
nose   26, 27, 33, 34, 35, 36

oxygen   16

pasteurisation   17
pathogens   3, 86–90
personal hygiene   2, 22, 25, 27, 28, 33–6, 54, 55, 80, 88, 89
pest control   64–73
pets   81
pH   12, 18, 43
potentially hazardous foods   11, 12, 22, 37, 39–42, 50, 51
precleaning   58

premises design 74–9
preparation of food 54, 77
preservatives 13, 19
properties of food 11–13
purchasing food 49

rats 65, 66–8
refrigerator 13, 14, 51–52, 54, 79, 81
refrigerated food 13, 14, 38, 51–52, 53, 54, 82
refuse disposal 79
reheating of food 14, 55, 56
*Salmonella* 5, 12, 21, 23–6, 36, 37, 38, 39–42, 53, 69, 86
salt 18
sanitisers (see disinfectants) 2, 59
sanitising 57
sanitising schedule, cleaning and 59–63
self-service 55
serving food 55
*Shigella* 86, 88
sinks 78, 79, 82
smoking
  cigarette 27, 35–6
  of food 18
spices 18
spoilage of food 3
spores
  bacterial 7–8, 12, 28, 29, 30, 36, 37, 38, 39, 41, 42
  mould 7, 10
*Staphylococcus aureus* 4, 21, 23, 26–27, 35, 37, 38, 39–42, 86
structural conditions of food premises 76–9
storage
  of food 50
  of sanitised equipment 59
stored food insect pests 65, 71–73
sugar 18, 43
supervision 46, 47, 63
symptom of food poisoning 86–87

tea towels 54, 59, 83
temperature
  effect on growth 13–14, 17, 22, 53, 58, 80

freezer 14, 17, 52–4, 82
hot food 13, 55
refrigeration 13–14
transport of food 80–81
thawing 53, 80
thorough cooling
through cooking 54
time 14, 22
tissue 36
toxin 8
  from *Bacillus cereus* 23, 30
  from *Campylobacter* 88
  from *Clostridium botulinum* 89
  from *Clostridium perfringens* 23, 28
  from *E. coli* 88
  from *Shigella* 88
  from *Staph. aureus* 23, 26
  from *Yersinia* 89
  of ciguatera 84–5
  of mould 21

unconscious body habits 33, 35, 54
ultraviolet light 1

vegetable storage 51
vegetative cells 6–7
ventilation of kitchens 77
vibrio 4
*Vibrio parahaemolyticus* 4, 23, 31–2, 36, 37, 41, 86
viral gastroenteritis 21, 87
viruses 8
viral food poisoning 8–9, 21

walls of kitchens 77
washing
  by dishwasher 73
  by hand 58, 73
  of hands 73, 79
  temperatures
waste disposal 79
water activity 12, 17, 43

yeast 9
*Yersinia enterocolytica* 86, 89